New Ideas in Technique and Location

Directed explosions, hydro-radio and optical location

S. N. Dolya

This book is a collection of articles by a well-known physicist S. N. Dolya working at the Joint Institute for Nuclear Research in Dubna, Moscow region, Russia..

The articles show different aspects in radio, optical, infrared location of bodies, as well as an opportunity of determining coordinates of the bodies being underwater by means of scattering the electromagnetic waves over the acoustic waves. Using the two mirror-lenses infrared telescopes it is possible to determine the coordinates of a hot moving body. According to the Doppler frequency shift it is possible to reliably determine the coordinates of the body moving under the water. A satellite will detect the thermal footprint of the body moving under the water at a depth of 200 m. A comparison of the accuracy of determination (from the satellite) coordinates of the hot moving body by using an infrared telescope and a Doppler radar. It is shown that it is possible to register the ultra violet radiation reflected from the surface of the water by the matrix of micro pixel diodes located on satellite B in case if the water surface is irradiated with a narrow laser ray from satellite A. We discuss the structure of the equipment for registering a turbulent trace of the body both in the water and air.

The parameters of the underwater directed explosion have been calculated. It is shown that it is possible to control the flight path of the body, which does not have its own self-targeting system. The load capacity of the captive balloon has been estimated. An opportunity of accelerating the body till hypersonic velocities (5km/s) has been shown: it can be done by using the gas-dynamic method (not by a rocket).The conditions of microbe sterilization of large (10^4 km^2) areas have been considered on the surface of the Earth by means of artificial ozone holes. The conditions of coincidence of two bodies have been discussed for two cases: at almost ballistic motion of the bodies and in the case when the bodies are quickly maneuvering. The book discusses an opportunity of concentrating the energy of the explosion and transmitting this energy to a small physical body.

The book has been written for a wide circle of attentive readers who can find a lot of interesting information for themselves.

Content

Location

Technique

Location

Detection of underwater objects

Imagine at some point in the ocean there is an overwater or underwater vehicle that emits a sonic wave with a power $I_{irr} = 5 * 10^5$ W / m², at a frequency $f_{sonic} = 10$ kHz. This radiation spreads in all directions and at a distance $L_{det} = 30$ km from the ship it creates a sonic pressure of the order of $p_1 = 17$ W / m². The sonic wave reflected from the underwater object reflectance $k_{refl} = 10^{-2}$, produces a diffraction grid corresponding to the cylindrical sonic wave due to the compressibility of the water. High-frequency generators, with a capacity of $P_{gen} = 500$ MW, operating at a frequency of $f_{radio} = 10^8$ Hz, irradiate some parts of the surface of the water with a narrow beam of radio waves. Due to the fact that the scattering takes place on the traveling grid, the electromagnetic wave reflected from it, is Doppler shifted by the value of $\delta f = 100$ Hz. The coordinates of the underwater object are determined by the registered diffraction radiation.

Introduction

The physical value characterizing the sound - is intensity I. It is measured in units: W / m². For a flat sinusoidal wave we have the following ratio:

$$I = p^2 / 2\rho c, \tag{1}$$

where: p - the amplitude of the sound pressure, ρ - density of the medium, c - speed of the sound in the medium. For the sea water we have: $\rho = 10^3$ kg/m³, c = 1.5 km /s.

Conventional emitters have the intensity [1], p. 119, the order of 10^5 W / m², and according to the formula (1) the sonic pressure level is equal to:

$$p_1 = (I * 2\rho c)^{1/2} = 5 * 10^5 \text{ Pa.} \tag{2}$$

Let the emitter have a radius of 1 m, the wave from the transmitter is spherical, the pressure therein weakens as 1 / r, at a distance of 30 km the sonic pressure in the sonic wave decreases by $3 * 10^4$ times. For the frequency of the sound wave $f_{sonic} = 10$ kHz pressure has decreased by 10 times due to additional wave attenuation [2], so that the pressure of the incident wave will be approximately equal to $p_2 = 1.7$ Pa.

When the reflection coefficient of the sonic wave from the object $k_{refl} = 10^{-2}$, then the pressure in the reflected wave will be as follows: $p_3 = 1.7 * 10^{-2}$ Pa. After returning to the starting point, the pressure in the sonic wave will fall again by $3 * 10^4$ times and taking into account the attenuation, it will be equal to $p_4 = 5 * 10^{-8}$ Pa, which is impossible to detect.

The sensitivity of the best hydrophones [3] is of the order of 10^{-4} V / Pa, so that the pressure of the order of p_4 is unavailable for them.

The minimum sonic pressure which can be comprehended by normal people aged from 18 to 25 years is of the order of [4], 10^{-5} Pa. Electro-acoustic transformers (hydrophones) have not yet reached the sensitivity of the people. It can be seen that the pressure P_4 is much less than $p_0 \approx 10^{-3}$ Pa. It is the minimum pressure in the sonic wave which may be registered by technique.

1. Physical motivation of the task

The sonic waves in the water can spread out for relatively long distances [2]. The lower is the frequency of the sound, the longer distances it spreads out. Reduction of the pressure in the sonic wave in this case is defined by its attenuation. While propagates it "captures" more and more space.

Assume that there is a ship in the ocean which emits a sonic wave with a frequency f_{sonic}, producing a sonic field around itself. Suppose that the sonic frequency is: $f_{sonic} = 10$ kHz, then the length of the sonic wave is equal to: $\Lambda_{sonic} = c / f_{sonic} = 15$ cm. Let the underwater object of the cylindrical shape be approximately at a depth $h_1 = 400$ m from the surface.

Scattering of the flat sonic wave on the cylinder is well known [5]. The most efficient scattering is when the wavelength of sound is approximately equal to the perimeter Λ_{sonic} of the cylinder. In this case, the intensity of the scattered wave is approximately 2 times less than the incident one. The polar scattering diagrams are given in [5]. For the short wavelength, in comparison with the perimeter of the object, a sufficiently large incident wave power dissipates in the opposite hemisphere [5].

Due to the compressibility of the water the pressure of the sonic wave will produce periodic changes in density and, finally - periodic changes in the relative dielectric permittivity ε.

2. Schemes of electric equipment

Figure 1 shows a diagram of complex equipment. The ship (1) located in the ocean, by means of a specialized emitter (2), generates sonic waves (3). Scattering on the underwater objects (4), the waves are transformed into the cylindrical ones (5). These sonic waves form a diffraction grid (6). Aircrafts (7), which have emitters of electromagnetic waves, irradiate different sites of the water surface by radio waves (8). The radio waves (9) scattered on the grid produced by the sonic wave, are recorded by the receivers located on another group of aircrafts (10).

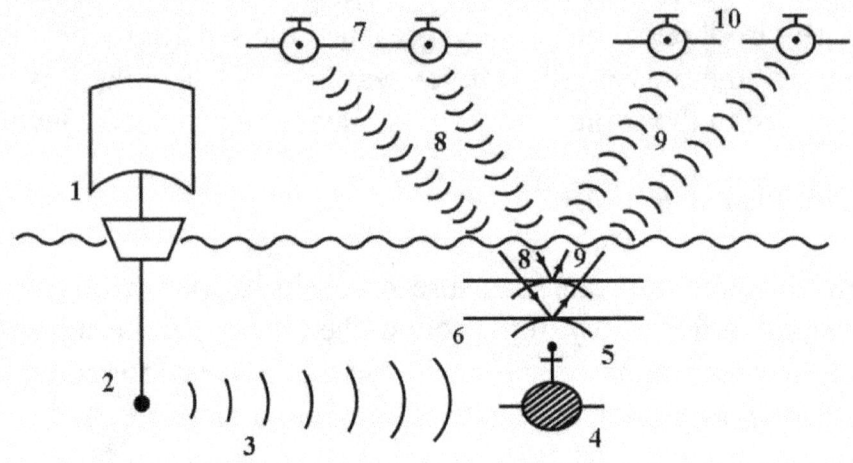

Fig.1. Scheme of the equipment

2. 1. *Reflection of radio waves from the water surface*

Fresnel's formulae [1], p. 833, relate the reflection coefficient of the electromagnetic wave power with the parameters of the medium.
For normal incidence, we have [1],

$$r_{refl} = (n_2 - n_1)^2 / (n_2 + n_1)^2 \qquad (3)$$

wherein n_1 is the breaking index of the first medium, n_2 - the breaking index of the second medium. In our case, $n_1 = 1$ is the air, $n_2 \approx 10$ - water [6].

Substituting numbers into the formula (3), we find that due to the difference between the breaking indices of the air and water, in the case of normal incidence, only 1/3 of the incident wave power will penetrate into the water. The same power losses will take place when radio waves are going out from the water into the air. It is explained by the following: the reflection coefficient does not depend on the side of the interface from which the incident radio waves come out.

2. 2. Radio waves breaking

Assume the incidence angle of the radio waves on the water surface with the normal to be equal to 30^0. Then due to the difference between the breaking indices of the air and water, the broken ray will propagate at an angle to the normal sine by 10 times less than the $\sin 30^0 = 0.5$. This angle with the normal will be equal to 3^0.

We have chosen the wavelength of radio waves $\lambda_{radio} = 3$ m, respectively, the frequency of the radio waves will be equal to: $f_{radio} = 10^8$ Hz. Due to a tenfold difference in the breaking indices of the air and water, the wavelength of radio waves in the water will be equal to: $\lambda_{radio1} = 0.3$ m.

2. 3. Conditions for Bragg reflection.

It is easy to see that in this case the condition of Bragg reflection for radio waves from the sonic waves:

$$2\Lambda_{sonic} * \sin\Theta_B = \lambda_{radio1}, \qquad (4)$$

where $\Theta_B = 87^0$ - angle of incidence of the radio waves onto the sonic wave, $\sin\Theta_B \approx 1$, $\Lambda_{sonic} = 15$ cm, $\lambda_{radio1} = 30$ cm.

The same will be the reflection angle of the radio waves from the sonic waves. Coming out of the water into the air, these radio waves will propagate at an angle of 30^0 to the normal and can be detected by the receiver.

2. 4. The intensity of the Bragg reflection

The formula for the intensity of the Bragg reflections in the first order is well known [7]:

$$I_1 = I_{00} * \sin^2 (\pi p n^3 S_0 L / 4\lambda_{radio}), \qquad (5)$$

where I_{00} is the intensity of the incident radio waves, $p = (\varepsilon_0 - \varepsilon) / \varepsilon_0^2 S$, elastic-optical constant, ε_0 - the permittivity of the unperturbed medium, ε - the dielectric constant of the medium where the sonic wave propagates, S_0 is the strain amplitude in the sonic wave, L - the length of the acousto-optic interaction, n – the breaking index for the radio waves.

The dielectric constant of water at this frequency [6] is equal to: $\varepsilon_0 \approx 10^2$, dielectric constant ε of water, where the sonic wave propagates, can be determined from the following considerations.

For water the compressibility factor is [8], p. 71, $k_{water} = 5 * 10^{-10}$ Pa^{-1}. For the sonic pressure $P_{1\,Pa} = 1$ Pa $= 10$ dn / cm^2, at a frequency $f_{sonic} = 10$ kHz, it is possible to find the water particle velocity from the following ratio:

$$P_{1\,Pa} = 1\ Pa = \rho_{water} * V^2/2. \tag{6}$$

From the above we find that the rate of displacement of the particles in the sonic wave is equal to:

$$V = (2P_{1\,Pa} / \rho_{water}) = 4.5\ cm/s, \tag{7}$$

the corresponding shift S can be found from the relation:

$$V = S * 2\pi\ f_{sonic}, \tag{8}$$

and it is equal to: $S = V / 2\pi f_{sonic} = 7.1 * 10^{-5}$ cm. Displacement S_0, in the cylindrical sonic wave scattered on the underwater object under the pressure $p_3 = 1.7 * 10^{-2}$ Pa, is equal to $S_0 = 2.8 * 10^{-5}$ cm.

Now we can calculate the elastic optical constant $p = (\varepsilon_0 - \varepsilon) / \varepsilon_0^2 S$. For sonic pressure $P = 1$ Pa the density of water and the dielectric constant will differ from the value of ε_0 by the value of $5 * 10^{-10}$. The elastic optical constant for water, thus, can be estimated as follows:

$$p = (\varepsilon_0 - \varepsilon) / \varepsilon_0^2 S = 5 * 10^{-10} / (10^4 * 7.1 * 10^5) = 7 * 10^{-10}. \tag{9}$$

The value pS_0 in the argument of the sine is equal to: $pS_0 = 2 * 10^{-14}$.
For small values of the argument the function sin x can be substituted by x, and thus, the expression (4) for the intensity of radio waves in the first diffraction order can be written as follows:

$$I_1 = I_{00} * (\pi p n^3 S_0 L / 4\lambda_{radio})^2. \qquad (10)$$

From this formula it is clear that the intensity of the scattered radio waves is very small in comparison with its incident wave intensity on the surface of the water.

2. 5. *The depth of penetration of radio waves into the sea water*

We estimate the value of L - the length of the interaction of radio waves with the sound, in fact the depth of penetration of the radio wave into seawater.

The conductivity of the seawater in units $(Ohm * m)^{-1}$ [8], p.1000 is equal to $5 * 10^{-2}$. This means that the electrical resistivity of seawater at a salinity of 4 ppm (4 grams of salt per liter) power is $\rho_{sea\ water} = 2 * 10^3\ Ohm * cm$ and its conductivity in units of $1 / c$, is equal to:

$$\sigma_{sea\ water} = 9 * 10^{11} / \rho_{sea\ water} = 4.5 * 10^8\ 1 / s. \qquad (11)$$

Now we can calculate the depth of the skin layer.

$$L = \delta_{sea\ water} = c / 2\pi (\sigma_{sea\ water} * f_{radio}) = 0.22\ m. \qquad (12)$$

Thus, the ratio $L / \lambda_{radio} = 7 * 10^{-2}$, i.e., the depth of penetration of radio waves in the salt water at this frequency is approximately 0.1 of the wavelength in vacuum. For the case of fresh water due to absorption of the radio wave by water molecules, the depth of penetration of the composition [6] will be several wavelengths.

This small ratio of the diffracted power to the incident one can be explained by two reasons. The first one is a very small intensity of the sonic waves reflected from the object underwater. The second reason is a small area of interaction of the radio emission from the scattered sonic waves.
We calculate the intensity of the diffracted radio waves by using formula (10):

$$I_1 = I_{00}*(\pi p n^3 S_0 L /4\lambda_{radio})^2 = I_{00}*(3.14*2*10^{-14}*10^3 *7*10^{-2} /4)^2 = I_{00}*10^{-24},$$

where $n \approx 10$, the breaking index of the water for this radio wave frequency [6].

It is interesting to note that in our case the intensity of the diffracted

9

radiation, calculated according to the formula Raman - Nat

$$I_1 = I_{00} * J_1^2 (\pi p n^3 S_0 L / 2\lambda_{radio}), \tag{13}$$

gives exactly the same value for the diffracted radiation, calculated according to the Bragg formula. This is due to the fact that the asymptotic expansion of the first order Bessel function at small values of the argument $J_1(x)$ is equal to $x/2$, and the asymptotic expansion of $\sin(x/2)$ is also equal to $x/2$. The double Bragg angle is the angle of deflection of the beam from its original direction, equal to the angle of diffraction in the case of Raman - Nat.

2. 6. Received power

Let the diameter of the radiating radio wave antenna is: $d_{ant} = 30$ m, so that the cone angle of the radio emission is: $\Theta_{rad} = \lambda_{radio} / d_{ant} = 0.1$. Then, at the height of the airplane $h_{plane} = 3$ km, the radio wave irradiated area will have a diameter of $d_{plot} = h_{plane} * \Theta_{rad} = 300$ m.

Let the pulsed power generator be $I_{radio} = 5 * 10^8$ W, the pulse duration $\tau_{pulse} = 10$ ms, the pulse repetition rate $F_{triod} = 10$ Hz. As we remember, only one third of the incident power will penetrate into the water and one third of the diffracted radiation will come out from the water. Then the diffracted power of the radio waves will be equal to the following:

$$I_{dif} = I_{radio} * 10^{-25}. \tag{14}$$

3. Sensitivity of the method

This method assumes that the detection of underwater objects requires that the aircrafts would irradiate the water surface and receive the reflected signals. It can be seen that the ship in the ocean can generate radiation occupying the space in the area of the order of $d_{inv} = 30$ km. The cylindrical sound wave most strongly manifests itself in the area with a side of approximately $d_{sq} = 300$ m. Thus, the total area under study can be represented as 10^4 of small squares; from each of them it is necessary to register the reflected radio waves during the flight of the aircraft.

Since the width of the beam of the radio emission is $\Theta_{rad} = 0.1$, then the distance between the emitting and receiving radio waves along the propagation direction is equal to: $S_{plane} = 6$ km, the broadening of this kind of the antenna beam will be $\Delta h_{radio} = \Theta_{rad} * S_{plane} = 600$ m. Let the receiving antenna have the same diameter as the radiating one, $d_{ant} = 30$ m, its area is $S_{ant} = 700$ m^2. This receiving antenna absorb $4S_{ant}/\pi * \Delta h^2_{radio} = 2 * 10^{-3}$ of the diffracted radiation.

Finally, the received power will be equal to:

$$I_{dif} = I_{radio} * 10^{-25} * 2 * 5 * 10^{-3} = 5*10^8 * 2 * 10^{-28} = 10^{-19} \text{ W.} \quad (15)$$

You can determine the minimum noise power of the receiver as: $\Delta W = kT\Delta f$, where k is the Boltzmann constant: $k = 1.38 * 10^{-23}$ J / degree for $T = 2$ degrees Kelvin (liquid helium) and $\Delta f_{cav} = 10^2$ Hz. Then we have: $\Delta W = 3 * 10^{-21}$ W, so that the receiver whose input circuit is a superconducting resonator, has small noise interference. It can be seen that in this case the power of the received signal (15) will exceed the noise power in the resonator.

To do the reception band narrower than $\Delta f = 100$ Hz is not practical because it corresponds to the pulse duration: $\tau_{rad} = 1 / \Delta f = 10$ ms. Let the time of watching an area be equal to $T_{plot} = 0.1$ s. Then, the total time of the survey of the entire region with dimensions of 30 * 30 km, consisting of 10^4 small areas, will be: $\tau_{total} = 10^4 * 0.1 = 10^3$ s, or about 20 minutes.

There won't be any problems with mirror reflection of radio waves from the water surface. The coefficient of the mirror reflection can be tens of percent that is approximately by 25 orders of magnitude stronger than the reflection of radio waves from the diffraction of the waves scattered on the underwater object. However, due to the diffraction at the traveling grid, the radio waves reflected from it will have the Doppler shift by the frequency of the sonic wave [6], that is, $\Delta f_{radio} = f_{sonic} = 10$ kHz. The bandwidth of the receiver $\Delta f_{cav} = 10^2$ Hz, there is no problem of interference from the mirror reflection.

Modern superconducting resonators have Q-factor of the order of $Q = 10^{10}$ [9]. This means that the private band of this resonator is hundredths of a Hertz, $\Delta f_{cav} = f_0 / Q$, and it needs a special extension of the band width to process these received signals fast enough. As for the registration without interference of the radiation shifted by 100 Hz, relative to the resonance frequency of the circuit, the problem seems to be technically quite solvable.

References

1. Physical Encyclopedic Dictionary, ed. Prokhorov, Moscow, Soviet

Encyclopedia, 1983, p. 120

2. http://www.akin.ru/spravka/s_ocean.htm

3. http:// www.zetms.ru/catalog/vibrodats/hydrophone.php

4. G. W. Kaye and T. H. Laby, Tables of physical and chemical constants, GIFML, Moscow, 1962, p. 73

5. http://corpuscul.net/teoriya-zvuka-2/rasseyanie-zvuka/

6. http://www.meteolab.ru/projects/dielectric/

7. http://www.femto.com.ua/articles/part_1/1076.html

8. Tables of physical quantities. Handbook ed. I. K. Kikoin, Moscow, Atomizdat, 1976

9. http://www.linearcollider.org/about/Publications/Reference-Design-Report

The complex of equipment for aerial observation

To raise any objects for aerial observation, it is necessary to use balloons. The balloons are held by steel cables with a cross section of 1 cm^2. To a height of h = 2 km they raise mirror-lens telescopes with a mirror diameter d_m = 0.7 m and mosaic photo detector devices. The observation angle of each telescope is of the order of 5 degrees. According to the known distance between the telescopes (the side of the triangle) and the two corners adjacent to this side, the vertex of the triangle is reconstructed. The velocity of the object is determined by the rate of its position change.

1. Detection Range

The range of direct vision ls can be found from the Pythagorean theorem if we imagine a right triangle in which one leg is R_E = 6400 km, i.e., - the radius of the Earth. The hypotenuse is equal to the R_E + h, where h is the location height of the observation point above the Earth. Then l_s, i.e., - the other leg of the triangle is $[(R_E + h)^2 - R_E^2]^{1/2}$, and for h $\ll$ R_E, we obtain the length of the direct observation.

To clarify what distances we are speaking about, it is necessary to substitute

12

the numbers into the above formula. Thus, for h = 2 km, the distance of the direct "vision" is found to be equal to l_s = 160 km, which is of interest for a number of applications. According to the known distance between the telescopes (the side of the triangle) and the two corners adjacent to this side, we define the vertex of the triangle, i.e., - the current vertex coordinates of the object. The velocity of the object is determined by the rate of change of its position.

2. Selection of the balloon parameters

We assume that the mass of the payload, which must be raised to a height of h = 2 km is 1 ton. According to the barometric formula, the density of air decreases with the height exponentially $\rho = \rho_0 \exp[-h / H_0]$, where h is the height of lifting, H_0 = 7 km, ρ_0 = 1.3 kg / m³ - the air density near the surface of the Earth. For the height h = 2 km the air density is ρ = 1 kg / m³. It is evident that in order to raise the weight of 1 ton to this height, you will need a hollow ball with a volume of about 10^3 m³. From the formula which relates the volume and diameter of the ball: $V_{b1} = \pi d_{b1}^3 / 6$, we find that the diameter of the ball d_{b1} should be about d_{b1} = 13 m.

The density of hydrogen filling the ball is 14 times less than the density of air, so that the diameter of d_{b1} = 13 m will have to be slightly increased.

We assume that the shell of the ball is made up of Mylar 25 microns thick. The Mylar density is 1.2 g / cm³, the volume of 1m squared Mylar is: S_m = 25 cm³, the weight of one Mylar squared meter is: m_m = 30 g.

Let the diameter of the ball be equal to: d_b = 20 m, then its capacity is: $V_b = \pi d_b 3 / 6$ = 4186 m³. The air density at the height h = 2 km is equal to ρ = 1 kg / m³. The load capacity of an empty ball will be equal to the following: $\rho * V_b \approx 4.2$ tons. Taking into account that the ball is not empty but filled with hydrogen, its load capacity at a height of h = 2 km is equal to 3.9 ton. Thus, the required mass of hydrogen inside the balloon is as follows: m_{h2} = 300 kg. The area of the shell of the ball, πd_b^2, is 1256 m², its mass m_{sb} = 40 kg. The weight of the steel cable with a cross section S_c = 1 cm² and the length h = 2 km, is equal to 1.6 tons.

Now it is necessary to solve the problem how to keep the ball at the predetermined height. When the diameter of the ball d_b = 20 m, the ball will "explode" upwards. The tensile strength of the steel rope is [1], p. 55

$\sigma_{st} = 340$ kg / mm^2. In comparison with nylon this value is $\sigma_{st} = 50$ kg / mm^2, but the density of nylon is about by 6 times less than that of the steel.

So, the aerostatic force acting on the cable is directed upwards and is 3.9 tons. The force of payload gravity equal to1ton and 1.6 ton weight of the cable are pulling the ball down to the Earth. It turns out to be the resultant upward force equal to 1.3 tons. The strength of the cable breaking is 34 tons, so that the accessible wind gusts going upward are 40 m / s. Indeed, the wind pressure is $P_W = C_x \rho V_w^2 / 2$. Considering that for our range of Reynolds numbers, Re $\approx 10^7$, and wind velocity V_w is ≈ 40 m / s, the drag coefficient of the ball, C_x is ≈ 0.5. Then we find that for wind velocity $V_w = 40$ m / s, the pressure will be as follows: $P_w = 10^3$ N / m^2, and the cross-section of the ball $S_{cb} = \pi d_b^2 / 4$ will be equal to 314 m^2, for the diameter of the ball d $_b = 20$ m. Then, the wind loading will result in additional tension of the cable with a force $F_w = P_w * S_{cb} = 31.4$ tons, if the wind is directed upwards.

Let the diameter of the bobbin on which the cable is wound be equal to $d_b = 3$m. Then the perimeter of one circle, the length of one turn of the wound cable on the bobbin, will be equal to P = 10 m. When winding one cable layer (200 turns), the height of the bobbin is equal to: $h_b = 2$ m.

3. Production of hydrogen

Consider the production of hydrogen to fill the ball by water electrolysis. As you know, one gram - mol of any gas under normal conditions occupies a volume of 22.4 liters. One gram mole contains 6 * 10^{23} molecules. The volume of 4.4 * 10^3 m^3 will occupy 10^{29} molecule, it is approximately 2 * 10^5 gram moles. The weight of one gram mole of hydrogen is 2 g, thus, the mass of hydrogen inside the ball will be equal to $m_{H2} = 2 * 2 * 10^5$ g = 400 kg. This value is by 30% higher than it is really needed: $m_{H2} = 300$ kg.

The mass of the water molecule is 9 times heavier than the mass of the hydrogen molecule. So, to obtain the mass of hydrogen of 300 kg, it is required to decompose 2.7 tons of water.

The first law of Faraday says that the mass of the substance released by the electrode is proportional to the current passing through the electrolyte and the current flow time through the electrolyte: m = k * I * τ. Since the hydrogen molecule is diatomic, then at the flow of the current in the I = 2 A for the time $\tau = 1$ s, 6 * 10^{18} molecules of hydrogen atoms will release on the cathode that on the charge it is one Coulomb.

Consider the generation of hydrogen by electrolysis of water. As you know, one gram - mol gases under normal conditions (we will not take into account that the density of gas at a height h = 2 km amounts to 30% less than under normal conditions) occupies a volume one gram mole cont.

The mass of the water molecule at 9 times the mass of the hydrogen molecule, so that to obtain an amount of hydrogen, m_{h2} = 300 kg, is required to decompose into oxygen and hydrogen is 2.7 tons of water.

To obtain 10^{29} molecules, it is required to extract the summed up charge of $2 * 10^{10}$ times greater. It is possible to reach by passing the current. I = 600 kA during time τ = 20 hours.

4. Electromagnetic waves radiated by an object

According to Planck's formula the body heated till 800 ^{0} K (~ 500 ^{0}C) has a maximum emission at wavelengths of 3-4 microns, which falls just within the band of atmospheric transparency. The spectral density of radiation is shown in the graph [2], p. 3, and it is equal to: 0.1 W / (cm^2 * str.* μ).

The transparency window occupies an area of 3.5 to 4 μ, for the mirror lens infrared telescope with a mirror diameter equal to: d_m = 0.7 m, the surface area of the mirror is equal to S_m = 0.3 m^2. The geometric factor (the solid angle – $S_m / 4\pi l_s^2$) in this case is 10^{-12}, and for the area of the radiation emission of hot gases equal to S_t = 2 m^2, we obtain the value of the received power equal to the following:

$$P = 0.1*2*10^4*0.5*10^{-12} = 10^2*10^{-11} \text{ W}.$$

Here we specify the factor of 10-11, because this is just the order of the threshold sensitivity D* of the infrared InSb detectors produced by company Hamamatsu, operating at the liquid nitrogen temperature [2], p. 9.

This means that the signal from the detector, for example, 1024*1024 pixels, with the total area of 1 cm^2, in the frequency band of Δf = 1 Hz, will exceed the noise by 2 orders of magnitude and be clearly detected.

5. Structure of infrared telescopes.

Let the diameter of the mirror be $d_m = 0.7$ m, and the radius of the sphere with the mirror bent on is equal to $r_m = 2$ m. Then, the mirror focal length is equal to $f_m = r_m / 2 = 1$ m. The focal length of the second lens is equal to 100 mm, and then the increase of the telescope will be 10 times more. There should be two telescopes of this type should to determine the distance to the target and see it stereoscopically.

It is known that, due to wave properties of light, the light wave can not be focused to the spot whose diameter is less than the wavelength. In our case the wavelength is equal to $\lambda_l \approx 4$ μ. That is why the matrix pixel of the mosaic photo detector must be of the same size. Taking into account the size of the space between the pixels and other supplementary guides, the total size of the matrix of 1024 x 1024 can be quite large - of the order of 6 x 6 cm. For this matrix the normal lens will have the focal length of the order of the diagonal size of the matrix, i.e. $f_{oc} = 100$ mm.

The angle of the normal lens lies in the range of 40-50^0. Then at the 10-fold increase of the telescope its observation angle will be 10 times smaller. This observation angle can be increased if to increase the focal length of the telescope ocular.

The telescope electric feeding can be carried out by using a high-frequency klystron, located on the ground [3]. Let the klystron radiation power be 2 kW continuously. At the efficiency of conversion equal to 50%, the power needed for devices, motors and appliances will be 1 kW - that is enough to control the rotation of the mirror, its inclination and the focal length of the telescope ocular.

Let the size of the emitting object is $l_t = 1.6$ m, it means that its angular size is $l_t / l_s = 10^{-5}$, and after the telescope ten fold increase the object will be seen as one hundredth of the solar disk, visible by the human eye. To correct the spherical aberration of such a telescope, you can use special lenses [4], and thereby - obtain an almost diffraction resolution.

6. Issues concerning the telescope stable position in the space

The application of this method depends on the telescope stable position in the space to have an opportunity for observing the low-flying targets.

The package of measures to stabilize the telescope can be divided into three parts: the aerodynamic stabilization of the captive balloon with a specialized wing located above the balloon pole, [5], the forced damping of the vibration to stabilize the locators placed on the tethered balloon [6], and, finally, the optical stabilization of the image [7].

7. Equipment complex operation

Figure 1 shows a diagram of the equipment complex for the aerial observation. At two movable platforms (1) there are arranged tethered balloons (2). The electric supply power is transmitted to the payload unit (5) by using the radiating antennas (3) and the receiving ones (4). Infra-red telescopes (6) are arranged in the blocks of the payload. The telescopes have an opportunity to rotate around the vertical axis and change their vertical angle due to their location on a horizontal shaft. It allows them to detect low flying objects (7).

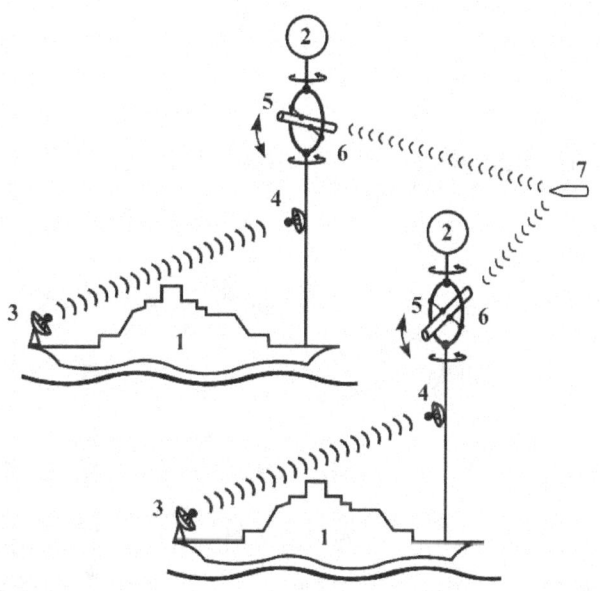

Fig. 1. Scheme of the equipment complex

Operation of the complex is carried out as follows. On two floating platforms (1) we switch on the electrolytic hydrogen with a capacity of 200 m³/ hour. During 20 hours the hydrogen fills the tethered balloons, (2), which rise to a height of h = 2 km. The balloons are held with steel cables of the total cross section $S_c = 1$ cm². By means of a radiating antenna (3) the complex obtains the energy by the receiving antenna (4). The balloons raise movable platforms (5) to a height h =2 km, where the infrared mirror-lens telescopes (6) with a mosaic photo receiver are located.

When operating in Omni direction regime the gyro stabilization is switched off. The platform rotates by 360 degrees around the vertical axis, and due to its position on the horizontal shaft the telescopes can change their angle towards the vertical axis. The observation angle of the telescope is of the order of 5 degrees. On the detection of a low-flying object (7) the gyro stabilization is switched on and two telescopes begin to follow the target.

The detection range limit is 160 km. The sensitivity D^* of the InSb detectors operating at the liquid nitrogen temperature, is better than $D^* < 10^{-11}$ W / (cm$^{1/2} *$ Hz). At the distance of 160 km from the radiator, having the temperature T = 500 ^{0}C and the area of 2 m^2 in the wavelength band from 3.5 microns to 4 microns the signal from the body exceeds the noises of the detector by 100 times.

References

1. Tables of physical quantities. Handbook ed. I. K. Kikoin, Moscow, Atomizdat, 1976

2. http://www.hamamatsu.com/resources/pdf/ssd/infrared_kird0001e05.pdf

3. http://ru.wikipedia.org/wiki/Беспроводная_передача_электричества

4. http://ru.wikipedia.org/wiki/Зеркально_линзовый_телескоп

5. V. A. Rozov, A. F. Konoha, I. A. Krendelev et al. Tethered balloon, resistant to wind load, RF Patent № 2279994

6. S.N. Sayapin, A.V. Sinev, A method for protecting an object on the suspended from the resonance oscillation and device for its implementation, RF Patent № 2245470

7. http://ru.wikipedia.org/wiki/Стабилизация_изображения

Detection of moving underwater objects

The article considers the reflection conditions of electromagnetic radiation with a wavelength of $\lambda_{radio} = 3$ cm from a soliton moving along the surface of the water. The soliton velocity can range from 15 - 30 m/s. It is shown that at the distance between the observer and the soliton equal to 30 - 50 km the power of the reflected radiation will be equal to the value of 10^{-22} from the initial power radiating the soliton. Thus, for the initial power $P_{rad} = 1$ kW the reflected power will be equal to 10^{-19} W, and this power can be reliably registered by modern electronic methods.

Introduction

Let a body be moving underwater towards the observer. In this case, then on the surface of the water there will b a small moving "bulge", this solitary wave is called a soliton.

The soliton velocity can range from 15 to 30 m / s. The distance from the observer to the soliton can be of the order of $L_{obs} = 30 - 50$ km. The physical task is to register this motion of the soliton by means of the Doppler frequency shift of the signal reflected from the soliton.

1. Physical motivation of the task

We have chosen the wavelength irradiating the surface water to be equal to $\lambda_{irr} = 3$ cm, the corresponding the wavelength to the frequency is equal to: $f_{irr} = c / \lambda_{irr} = 10^{10}$ Hz. The breaking index of the water for this frequency is $n_w = 3$ [1].

For the observer to have an opportunity to see the surface of the water at such a distance ($L_{obs} \sim 80$ km), he should be placed at a height above the surface of the water of more than $H_{obs} > L^2_{obs} / 2R_E = 0.5$ km, where $R_E = 6400$ km - radius of the Earth.

The reflection coefficient of electromagnetic waves from the interface between two media is determined by the Fresnel formulas [2]:

$$R_s = -\sin(\varphi - \varphi'')/\sin(\varphi + \varphi'') A_s$$

$$R_p = tg(\varphi - \varphi'')/tg(\varphi + \varphi'') A_p, \qquad (1)$$

where: φ - the angle of incidence of the electromagnetic wave (with the normal), φ "- the angle of breaking of electromagnetic waves, s and p - two different polarization of the wave. Fresnel formulas include the same law of reflection:

$$\varphi = \varphi' \tag{2}$$

where φ '- the angle of reflection (with the normal), and the law of refraction:

$$n_{air} \sin \varphi = n_w \sin \varphi", \tag{3}$$

where $n_1 = 1$ – breaking index of air, $n_w = 3$ – breaking index of the water, for a given frequency $f_{irr} = 10^{10}$ Hz.

As it can be seen from the scheme the angle of incidence is equal to $\varphi = \pi / 2$, from the relation (3) we can be find that: $n_w * \sin \varphi " = 1$ and $\sin \varphi" = 0.3$. Where we can find $\varphi " = 20^0$. Then

$$R_s/A_s = -\sin 70^0/\sin 110^0 = 1$$

$$(R_p/A_p)tg\ 70^0/tg\ 110^0 = 1. \tag{4}$$

The reflection coefficient for the intensity of the wave is as follows:
$R = \frac{1}{2} *[(R_s/A_s)^2 + (R_p/A_p)^2] \approx 1$.

This intensity of the electromagnetic wave is reflected forward. As for the back reflection by 180^0, this intensity can be estimated from the following considerations.

According to Lambert's law, the diffuse scattering intensity at different angles $I (\Theta)$ is related to the scattering intensity at zero angle $I(0)$ with the following ratio:

$$I (\Theta) = I(0)*cos\Theta. \tag{5}$$

For the angle $\Theta = (\pi / 2 - \delta)$ the cosine of this angle can be substituted by $\sin\delta$, and the sine for small angles can be replaces by the value of the argument. In this case, we get the following:

$$I (\Theta) = I_0 * \delta$$

Repeating this procedure once again, we obtain:

$$I(180^0) = I(0) * \delta^2, \qquad\qquad (6)$$

where $I(0) = I_{fal}$ which is the power reflected forward from the water surface, $\delta \approx 10^{-2}$ is the angle of incidence (from the horizon) of the electromagnetic power to the water surface.

Thus, the first coefficient in the ratio between the incident and reflected power is equal to the following:

$$G_1 = \delta^2 = 10^{-4}. \qquad\qquad (7)$$

The second coefficient is related to the aperture of the receivers and is equal to $G_2 = S_{ant} / 4\pi R^2$, that is the ratio of the area of the receiving antenna to the area of the sphere of the corresponding radius. Let us "smear" the diffusely reflected power over the surface of the sphere with radius R. Assume the receiving antenna area to be equal to $S_{ant} = 10$ m^2. The area of the sphere of radius R is equal $4\pi R^2 = 10 * 40 * 50 * 2 = 10^3$ km^2. Thus, the geometric coefficient G_2 in this case is equal to the following:

$$G_2 = S_{ant}/4\pi R^2 = 5*10^{-10}. \qquad\qquad (8)$$

Suppose that the observer has to view the surface of the water as the ring shape with the outer radius $R_{out} = 50$ km, and with the inner radius $R_{in} = 30$ km. Then the area of the ring is equal to: $S_{ring} \approx \pi (R_{out} + R_{in}) * (R_{out} - R_{in}) = 500$ km^2. Let the observer be looking at the areas with the size of 0.5 * 1 km. Then the area of one area is $S_{area} = 0.5$ km^2.

In the visible ring there are $S_{ring} / S_{area} = 10^3$ of these areas. Let the observer be looking at all the areas during a period of time $\tau_{obs} = 10$ s, then the viewing of one area takes the time equal to $\tau_{obs1} = 10$ ms.

The reflecting square of the soliton is assumed to be $S_{sol} = 10^{-2}$ m^2, then the ratio of the area of the soliton to the area of the area being viewed the will give another coefficient - $G_3 = S_{sol} / S_{area}$. In our case, this coefficient is equal to:

$$G_3 = S_{sol}/ S_{area} = 2*10^{-8}. \qquad\qquad (9)$$

Let the velocity of the soliton V_{sol}, (moving towards the observer), is

V_{sol} = 15 - 30 m / s. The Doppler-shifted frequency is higher than the frequency of the irradiating signal and the frequency shift is equal to:
Δf_{Dop} / f_{irr} = V_{sol} / c = 5 * 10^{-8} - 10^{-7}, where c = 3 * 10^{10} cm / s - the velocity of light in vacuum. In absolute terms this frequency shift Δf_{Dop}, for the chosen parameters will be equal to: Δf_{Dop} = 0.5 - 1 kHz. It can easily be detected by various methods.

2. Registration of the reflected radiation

We will consider a sequence of 10 superconducting resonators with quality factor Q = 2 * 10^8 and bandwidth Δf_{res} = 50 Hz. In order to clarify whether there is a signal with a corresponding shifted frequency, and what kind of frequency the incoming signal has, it is necessary to excite each of the 10 resonators by the incoming signal. Thus, the signal per one resonator is less 10 times than the incoming signal, the coefficient G_4 shows by how many times the signal per each resonator is smaller than the incoming signal. In this case G4 will be equal to the following:

$$G_4 = 10^{-1}. \tag{10}$$

Multiplying all the reducing coefficients of the received power in comparison with the incident power, we find that

$$G_1 * G_2 * G_3 * G_4 = 10^{-22}. \tag{11}$$

This means that if the power falling on one area is P_{irr} = 1 kW, then the received power is equal to P_{resiv} = 10^{-19} W. This power can be detected by modern equipment.

If we take a superconducting resonator, the intensity oscillations in it caused by thermal noises will be equal to:

$$P_{noi} = kT\Delta f, \tag{12}$$

where k = 1.38 * 10^{-23} J / degree - Boltzmann constant, T - temperature of the resonator in Kelvin scale, Δf = 50 Hz, is the frequency band. For temperature T = 1.4 K^0 the eigen value noise of the resonator will be as follows:

$$P_{noi} = kT\Delta f = 10^{-21} \text{ W.} \tag{13}$$

3. Signal accumulation

At the excitation of high-Q resonator by the harmonic signal, the amplitude of the oscillations in the resonator increases linearly up to values of the order $P_{resQ} = P_{irr} * Q$. For the selected parameters the power signal will correspond to the signal power in the non resonant case, which equal to the following:

$$P_{resQ} = P_{irr}*Q = 10^{-19}*10^8 = 10^{-11} \text{ W}. \tag{14}$$

That is 10 orders of magnitude higher than the noise power of the resonator.

Literature

1. http://www.meteolab.ru/projects/dielectric/

2. Tables of Physical Data, The Handbook, edited by I. S. Grigor'ev and E. Z. Meylikhov, Moscow, Nuclear Power Published, 1991

3. http://www.linearcollider.org/about/Publications/Reference-Design-Report, http://en.wikipedia.org/wiki/Superconducting_radio_frequency

Parameters of the thermal trace of a hot body moving under water

This article considers the problem of thermal parameters of a trace of a body moving under water at a depth of $y_0 = 200$ m, at a speed of $V_0 = 10$ m / s, and having the power of heat $P = 130$ MW. The additional temperature associated with the heat-generating body is equal to 10^{-6} degree. The heat trace has a width of 1 km and a length of 10 km. We discuss an opportunity of detecting this trace by means of the infrared telescope located at a height of $H = 300$ km. It is shown that to extract the signal from the background, the sensitivity of infrared detectors is required to be better than $D *> 3 * 10^{11}$ cm * $Hz^{1/2}$/ W.

1. Introduction. Motivation of the problem

Suppose, that in the semi-space $z > 0$, filled with water along axis z, at time $t = 0$ there is a body which begins moving at a constant speed $V_0 = 10$ m / s. Since the body is heated it leaves a heated water trace behind itself. The trace has a diameter $d_0 = 2$ m and it is heated to the temperature $T_1 = 21$ C^0. The outside medium temperature $T_0 = 20$ C_0. The water surface is located higher than the axis z being at the depth $y_0 = 200$ m.

The heat equation for this case is as follows:

$$\partial T/\partial t = a \{(1/r) \, d/dr \, [r \, dT/dr] + (1/r^2) \, \partial^2 T/\partial \varphi^2\} + F, \qquad (1)$$

where F - function of heat sources, i.e., in our case, - the heated water trace.

The value of the dimension coefficient $a = 1.4 * 10^{-7} \, m^2 / s$. Up to the radius $r < y_0$ the problem has a cylindrical symmetry.

The coefficient a is the ratio of the thermal conductivity of water $\lambda = 0.6 \, W / (m * degree)$, to the specific heat of water $C_{water} = 4.2 \, J / (g * degree)$ and its density $\rho_{water} = 1 \, g / cm^3$. Substituting the numbers, $a = \lambda / (C_{water} \, \rho_{water})$ [1], we find that

$$a = \lambda / (C_{water} \, \rho_{water}) = 0.6/4.2 = 1.4*10^{-7} \, m^2/s.$$

On the other hand, for water the coefficient of thermal conductivity is equal to [2], $\lambda = C_{water} * \rho_{water} * u_s \bar{L}$, where $u_s = 1.4 * 10^3 \, m / s$ - the speed of the sound in water, $\bar{L} = 10^{-10} \, m$ - the average distance between the water molecules. From the above formula we find that the value of a, calculated for these considerations, has the same value as well:
$a = \lambda / (C_{water} \, \rho_{water}) = u_s \bar{L} = 1.4 * 10^{-7} \, m^2 / s.$

The problem is to find the distribution of temperature T on the surface of the water: $T (x, y_0, z, t)$. Only the established process- when $z, t \to \infty$ is of our interest. It is clear that the temperature distribution $T (x, z)$ will move along the axis z with velocity V_0. It will look like as if the temperature distribution would be similar to the ridge of the finite length pulled by means of the rope.

2. Models of water heating

2.1. Model of the uniformly heated water disk

Let us calculate the temperature of the heated water disc having a height of $h_{disc} = 10 \, m$ and radius $r_1 = y_0 = 200 \, m$. The heat capacity of water is assumed to be equal to the following: $C_{water} = 4.2 \, MJ / (ton * degree)$. From the initial conditions we find that the body preserves leaving the heated cylinder trace with an area of $3.14 \, m^2$, with a length increasing per each second by 10 m. This trace is heated further up to $\Delta T_0 = 1$ degree.

Then, the energy yield from the body per second is as follows:

$$\Delta Q = 4.2 \text{ MJ} * 3.14 * 10 = 132 \text{ MJ},$$

i.e., the power of the energy yield of the body will be $W_{body} = 132$ MW.

The heated water disk with a height of $h_{disc} = 10$ m and radius $r_1 = y_0 = 200$ m has an area of $S_{disc} = 1.25 * 10^5$ m^2, volume $V_{disc} = 1.25 * 10^6$ m^3, and the yield of its energy $\Delta Q_{body} = 132$ MJ, the disc will be heated up additionally by $2.5 * 10^{-5}$ degrees.

By increasing the radius of the disk by 2 times till the value of $r_2 = 2y_0 = 400$ m, the temperature falls down by 2.7 times. This can be determined from the following relationships. When increasing the radius of the disk by 2 times the temperature would have dropped by 4 times, but taking into account that there is an air segment (it occupies approximately one third of the area of the circle), it is not heated and that is why the temperature will fall down only by 2.7 times and not by 4.

Now we know how the temperature decreases with increasing the radius of the disk. Then we can find the temperature distribution in the transverse direction x on the surface of water. Having the temperature T (r) we can calculate the value of the temperature T (ξ), where $\xi = \Delta x = \pm (r^2 - y_0^2)^{1/2}$. Thus, the value of r = 200 m, corresponds to $\Delta x = 0$, the temperature of r = 400 m corresponds to the temperature of $\Delta x = \pm 350$ m, and so on.

If we assume that the temperature reduction factor of 2.7 corresponds to zero temperature, the temperature distribution in the transverse direction will be in the form of an isosceles triangle with a height equal to $\Delta T_1 = 2.5 * 10^{-5}$ degrees, the excess temperature at the distance $\Delta x = \pm 350$ m is equal to zero, i.e. $\Delta T = 0$.

Thus, we can assume approximately the width of the water heat trace left by the heat-emitting body moving under the water is equal to about the fourfold depth of immersion of the body.

2. 2. *Model of the disc heating at which the temperature decreases linearly over the disc radius*

We assume that the disk temperature decreases over the radius according to the following law: $dT / dr = 1 / T^2$, $T = T_1 * d_0 / 2r$.

This form of the temperature distribution along the radius seems to be more realistic. Then, at the same value of Q, the integral ($\int m * C * T_0 * 2\pi rdr$) gives the value $\Delta T_2 = \Delta T_1 * d_0 / y_0$. For the diameter of the heated water disc $d_0 = 2$ m, and the disk radius $y_0 = 200$ m, we obtain the value of the excess temperature: $\Delta T_2 = 2.5 * 10^{-7}$ degrees.

2. 3. *Model of heating the disk with the temperature decline according to the law $1 / r^{1/2}$*

The solution of the heat conductivity equation can be found by the Fourier method. In this case, it can be represented as multiplication of the function depending on the time and the function depending on the coordinates. In the case of the cylindrical symmetry, the functions which depend on the coordinates are Bessel functions, which in their turn, at large values of the argument, will have the asymptotic behavior: $J_v (x) \sim 1 / x^{1/2}$.

Thus, it can be expected that the temperature will fall exactly over the radius according to the following law: $T (r) = T_0 * (d_0 / 2r)^{1/2}$. In this case, the surface water temperature will be less than for a uniform temperature distribution in accordance with the following factor: $\Delta T_3 = \Delta T_1 * (2/3) (d_0 / 2y_0)^{1/2}$. Substituting the numbers, we obtain: $\Delta T_3 = 2.5 * 10^{-5} * (2/3) (2/400)^{1/2} = = 1.25 * 10^{-6}$ degrees.

Assume the additional water temperature on the surface related with the moving body to be equal to: $\Delta T = 10^{-6}$ degrees.

3. Opportunities to separate the signal from the background

The spectral density of the blackbody radiation power is described by Planck's formula [3]:

$$u (\omega, T) = \omega^2 * h\omega/\pi^2 c^3 [\exp(h\omega/kT)-1], \tag{2}$$

where all the symbols are common and self-explanatory.

According to Planck's formula we can plot dependences of the power density radiation on the wavelength for different temperatures of the body.

From this graph it can be seen that the maximum radiation for the temperature $T_0 = 20 \ C^0$ lies near the wavelength $\lambda_0 = 10 \ \mu$. The very meaning of the radiation power density at the radiation maximum: $u = 10^{-3} \ W / (cm^2 \ str * \mu)$, [4]. In the wavelength $\lambda_0 = 10 \ \mu$ there is a transparency window of the atmosphere, with a width of a few microns. In this region, the photon energy $h\omega = 1.2 * 10^{-3} \ eV$ is approximately equal to the value of $kT = 2.6 * 10^{-2} \ eV$, so that $(h\omega / kT) = 0.5$.

The infrared telescope, located at a height of $H = 300$ km above the surface of the water, with a surface mirror area of $\pi d^2{}_{tel} / 4 = 0.5 \ m^2$ (the plate diameter 0. 84 m) will have a geometric factor $\pi d^2{}_{tel} / (4 * H^2) = 5 * 10^{-12}$. The total obtainable power in the band of wavelengths $\Delta\lambda = 1 \ \mu$ from the water surface area of 100 km², is equal to:

$$P_1 = 10^{-3} * 5 * 10^{-12} * 10^{12} = 5 \ mW. \qquad (3)$$

The zooming of the telescope in this case should be about 30-fold. The area of 10 * 10 km seen from a height of $H = 300$ km at an angle approximately equal to: $14/300 = 4.6 * 10^{-2}$ rad, where 14 km is the size of the diagonal section. Let the mirror receiving the electromagnetic radiation is curved over $R_{mirr} = 0.84$ m, then its focal length is equal to: $F_{mirr} = R_{mirr} / 2 = 420$ mm. Let the focal length of the ocular be equal to $f_{ocul} = 14$ mm. Then the ratio of the focal lengths F_{mirr} / f_{ocul} will be equal to 30. If the size of the photo detector array is equal to 1 * 1 mm then the area of the size of 10 x 10 km will be projected onto this matrix.

We divide the host matrix by 100 pixels, the area of each pixel is then equal to: 100 * 100 μ^2, and it will receive the background power of the value $P_3 = P_2 * 10^{-2} = 50 \ \mu W$ / pixel. The wave limit of radiation focusing with a wavelength $\lambda = 10 \ \mu$ is equal to the wavelength, i.e., we may reduce the size of the pixel by 10 times more.

The very magnitude value of the background signal $P_3 = 50 \ \mu W$, is large enough. We are also interested in the opportunity to extract additional radiation related with the temperature rise by $\Delta T = 10^{-6}$ degrees in comparison with the neighboring areas.

Let the telescope receiving radiation move with velocity V_{tel} = 10 km / s. The distance of 1000 km will be covered by it during $\Delta\tau_{tel}$ = 100 s. While monitoring it has "to look through" a square with a side of 1000 km, i.e., this area will contain 10^4 sites with a dimension of 10 x 10 km. Then, the time of "viewing" one section should be not more than 10 ms, that corresponds to the frequency band of the detector equal to $\Delta f = 10^2$ Hz.

While the telescope covering the distance of 1000 km, the body will move forward for a distance of 1000 times less, i.e., - 1 km. If to monitor only this area, then the body will cover the distance of 1 pixel over the matrix. The ten-kilometer trace of the heated body will occupy 10 pixels on the matrix.

We expand the function u (ω, T) in a series on the degrees of T, and we limit ourselves with the first term:

$$u = u_0 + \Delta T * du/dT, \tag{4}$$

where u_0 corresponds to the background radiation value, $\Delta T = 10^{-6}$ degrees, the temperature rise above the background value.

Differentiation of the function u (ω, T) over T gives:

$$du/dT = u_0\, e^{h\omega/kT}\, {}^*h\omega/[kT^2\, {}^*(e^{h\omega/kT} -1)]. \tag{5}$$

Thus, for u we obtain the following expression:

$$u = u_0\{1+ e^{h\omega/kT}\, {}^*(h\omega/kT)^*(\Delta T/T)\, /(e^{h\omega/kT} -1)\}. \tag{6}$$

Substituting in the formula (6) the values of: (hω / kT) = 0.5, T = 293 K^0, $\Delta T = 10^{-6}$ K^0, we obtain:

$$u = u_0* (1 + 1.65*0.5*10^{-6}/293*0.65) = u_0 *(1 + 4.3*10^{-9}). \tag{7}$$

The sensitivity of modern MCT infrared detectors operating at the liquid nitrogen temperature [4], is of the order D * = 3 * 10^{10} cm * Hz$^{1/2}$/ W. The inverse D * is equal to the minimum power level that can be registered by the detector.

In this case, the value 1 / D * is 1 / D * = 3 * 10^{-11} W / cm * Hz$^{1/2}$. For the detector with an area of 10^{-4} cm^2, having a frequency band $\Delta f = 10^2$ Hz, the lowest sensitivity of the detector will be $P_{min} = 3 * 10^{-12}$ W.

The excess signal power P_4, in our case is equal to:

$$P_4 = P_3*4.3*10^{-9} = 2*10^{-13} \text{ W},\qquad(8)$$

it is smaller by- the order of magnitude.

4. Reducing the temperature of infrared detectors

The transition from the nitrogen temperature $T_{N2} = 77$ K^0 to the helium temperature $T_{He} = 1.8$ K^0 well used in the accelerator technology, will allow one to reduce the power of the thermal noise $kT\Delta f$ by about 40 times. It is necessary to mention that there are other types of noises beside the thermal ones. Nevertheless, the thermal noises of the detector will reduce by 40 times in this case.

Conclusion

Registering the area formed by the heat-emitting body moving under water is of interest because this thermal trace is much bigger than the body itself and that is why it is easier to be detected than the body.

References

1. Tables of physical quantities, Directory Ed. I. K. Kikoin, Moscow, Atomizdat, 1976

2. Physical Encyclopedic Dictionary, ed. A. M. Prokhorov, the thermal conductivity of the article, p. 748, Moscow, "Soviet encyclopedia", 1983

3. http://ru.wikipedia.org/wiki/Формула_Планка

4. http://www.hamamatsu.com/jp/en/product/category/3100/4007/index.html

A comparison of possibilities to measure the coordinates of a moving hot body using an infrared telescope or Doppler radar

A comparison of the sensitivities of methods which allow us to determine the coordinates of a moving hot body is made. The infrared telescope can reliably distinguish the signal from the body against the background of thermal radiation from the Earth's surface and reflected solar radiation. The accuracy of determination of the body's coordinates is of the order of 100 meters. The Doppler radar has an accuracy of coordinates' determination of about 10 kilometers and requires deployment in the Earth's orbit of a greater number of satellites compared to the infrared telescope.

Introduction

The infrared telescope and the Doppler radar are research instruments with a very high sensitivity. Three telescopes or radars of such type that are placed at the vertices of a triangle in which the Earth is "inscribed" will observe its total circumference. In order to view the entire surface of the Earth, three additional satellites rotating in the orthogonal plane ought to be used.

Let us assume that a hot body moves above the Earth's surface. We will compare then the sensitivities of the above two instruments in determining the body's coordinates.

For that purpose we will calculate the intensity of radiation received by the antenna (mirror) with a diameter of 4 m which is placed at a satellite rotating round a circular orbit at the distance $H_{sat} = 6400$ km from the Earth's surface. Such a satellite can view a region of order of about $10^4 * 10^4$ km^2 on the surface of the Earth.

Three satellites rotating in an equatorial orbit will view a ring with a width from the 0th to 60th parallel north and south, i.e. approximately from the South Polar Circle (67° of south latitude) to the Arctic Circle (67^0 of north latitude).

Three more satellites which move along a meridional orbit will observe then the polar regions. And a satellite above the pole will survey the area from the pole to the 30th parallel, which is almost up to the tropic located at the 23rd parallel.

I. Infrared telescope

1. Sensitivity of instruments

Planck's law gives the radiation intensity of a blackbody as a function of the temperature. A curve for the spectral intensity of the radiation flux against the body's temperature is depicted in [1]. It can be determined from the curve that for the temperature $T = 800$ K^0 the radiation power equals:

$$W_{infrared} = 10^{-1} \ (W^* \ str^{-1} \ cm^{-2} \ \mu^{-1}). \tag{1}$$

The radiation peak is then in the wavelength range of 3-4 microns. Shown in [1] is also a curve of sensitivity of various infrared detectors. From this curve it follows that the InSb (-196 C^0) detector is suitable one for this wavelength region, and its sensitivity here is: $D^* = 10^{11}$ (cm * Hz$^{1/2}$/W).

We define now the geometric factor G of the receiving instruments as the ratio of the IR telescope (12.56 m^2) to the square of the radius. Then $G = 12.56 \ m^2/4\pi H_{sat}^2 = 2.4 * 10^{-14}$. The value $4\pi G$, which is the spatial angle occupied by the telescope mirror, is: $4\pi * 2.4 * 10^{-14} = 3 * 10^{-13}$ str.

We assume that the radiator's area $S_{rad} = 3 * 10^4$ cm^2. Then the power coming from the body to the telescope mirror is:

$$W_{infrared} = 10^{-1}*4\pi *2.4*10^{-14}*3*10^4 = 9*10^{-10} \ W.$$

The matrix of the above photo detector must contain $10^5 * 10^5$ pixels and occupy an area of 1m * 1m. Here, each pixel corresponds to a region on the Earth's surface with dimensions 100 m * 100 m.

We determine now the required bandwidth of this detector. A satellite will travel with a speed of 10 km /s, passing through one pixel in the matrix during 10^{-2} s. The reciprocal of 10^{-2} s, i.e. 10^2 Hz, is the required bandwidth of a receiving photo detector in the matrix. The expression for the sensitivity of this detector will contain this value as a square root, that is the numerical value of this quantity is 10 Hz$^{1/2}$.

Let us calculate next the area of one photo detector in the matrix. The entire matrix contains about $10^5 * 10^5$ pixels and occupies an area of 1 * 1 m. Then the size of one pixel is 7 * 7 μ.

One cannot use here a photo detector of smaller size because radiation cannot be focused into a region smaller than the wavelength. In our case, the largest wavelength of radiation is: $\lambda = 4\ \mu$. So, we choose the size of one pixel as $7 * 10^{-4} * 7 * 10^{-4}$ cm^2.

The expression for the sensitivity of the detector includes this value also as a square root of the size. Therefore, the actual sensitivity of the detector, $1 / D^*$, is: $1 / D^* = 10^{-11}$ [W / (cm * Hz$^{1/2}$)] $= 10^{-11} * 7*10^{-4} * 10 = 7 * 10^{-14}$ W.

This is the threshold sensitivity of the detector, $1 / D^* = 7 * 10^{-14}$ W. It is four orders of magnitude smaller than the value of the signal: $W_{infrared} = 9 * 10^{-10}$ W. Thus, it can be seen that the detector sensitivity is sufficient as the radiation is focused here to one pixel.

2. Background conditions. Thermal radiation from the Earth's surface

Each pixel in our case corresponds to a region of size 100 * 100 m where a hot radiating body is located. This body is "hot", with $T = 800$ K^0, but its surface area is only 3 m^2. Next, we assume that the Earth's surface over which this body is found has a temperature of 273 K^0 (0 degrees Celsius), and its area is 100 m^2.

The intensity of radiation emitted by a body with $T = 273$ K^0 in the wavelength range 3-4 micron is 10^{-5} (W / str * cm^2 * μ), i.e. four orders of magnitude lower than the radiation from a body with $T = 800$ K^0. The ratio of the surface areas of the body and the Earth is 3 m^2/10^4 m$^2 = 3 * 10^{-4}$.

This means that the signal here will be three times greater than the background.

3. Background conditions. Reflected solar radiation

The power of solar radiation near the Earth's surface is of order 1 kW/m^2. We assume that such power is uniformly distributed over the spectral range with a width of 100 μ. This denotes that the spectral range of width 1 μ (from 3 to 4 μ) will account for an incident power of 10 W/m^2.

Let us take the coefficient of reflection of solar radiation from the Earth's surface as 30%. We obtain then that the intensity of reflected radiation falling on one steradian is: $P_{refl} = 10 * 10^4 * 0.3/4\pi = 2.4 * 10^3$ W * str^{-1}.

The power $P_{refl} = 2.4 * 10^3$ W * str^{-1} is of the same order as the radiation power of a hot body, $W_{infrared} = 10^{-1} * 3 * 10^4 = 3 * 10^3$ W * str^{-1}. This suggests that in the areas with direct incident and reflected solar radiation, the operation of an infrared telescope will be inefficient. Such a telescope can be effective only where the solar radiation is slanting or absent.

II. Doppler radar

We will now calculate the intensity of reflected microwave radiation falling on a spherical mirror with a diameter of 4 m which is placed on a satellite of the same kind.

1. Equipment

Let us choose a wavelength for the radiation that will be emitted onto the Earth's surface from a satellite: $\lambda_{irr} = 0.4$ cm, which corresponds to the wave frequency $f_{rad} = c / \lambda_{irr} = 7.5 * 10^{10}$ Hz.

We assumes that the first coefficient G_1 in the ratio between the power of the incident and reflected waves equals:

$$G_1 = 0.5. \tag{2}$$

The second factor is related to the aperture of the receiving apparatus: $G_2 = S_{ant}/4\pi H_{sat}^2$. The geometric factor G_2 is thus:

$$G_2 = S_{ant}/4\pi H_{sat}^2 = 2.4*10^{-14}. \tag{3}$$

A satellite can view a region of about $10^4 * 10^4$ km^2. The area of such a visible region is: $S_{reg} \approx 2 * 10^8$ km^2. Assume that the satellite observes one by one smaller plots with the dimensions 10 x 10 km^2. Then the area of each plot $S_{plot} = 100$ km^2 and the number of such viewed plots $S_{reg} / S_{plot} = 2 * 10^6$.

Assume also that the surface area of the body moving above the Earth's surface $S_{body} = 1$ m^2, then the ratio of the body's surface area to the area of the viewed plot gives another factor: $G_3 = S_{body} / S_{plot}$. In our case, this ratio is:

$$G_3 = S_{body} / S_{plot} = 10^{-8}. \tag{4}$$

Let the body's velocity V_{body} be equal to 0.3 - 3 km / s. The Doppler frequency shift $\Delta f_{Dop} / f_{rad} = V_{body} / c = 10^{-6} - 10^{-5}$, where $c = 3 * 10^{10}$ cm / s is the speed of light in vacuum. In absolute values such frequency shift Δf_{Dop} for the selected parameters equals 75 - 750 kHz. It can be easily detected by various methods.

We consider now a sequence of 100 superconducting cavities [2] with a Q-factor of 10^7 and a bandwidth Δf_{cav} of 7.5 kHz. In order to find out whether there is a signal with a frequency corresponding to shift one and which frequency has the incoming signal, each of the 100 cavities needs to be excited by a signal. A signal due to one cavity will be 100 times smaller than the whole incoming signal. The coefficient G_4 in this case is as follows:

$$G_4 = 10^{-2}. \tag{5}$$

By multiplying all of the above coefficients, we obtain:

$$G_1 * G_2 * G_3 * G_4 = 0.5*2.4*10^{-14}*10^{-8}*10^{-2} = 10^{-24}. \tag{6}$$

This denotes that if the incident power $P_{rad} = 1$ kW, then the received power $P_{rec} = 10^{-21}$ W. Such power can be detected with modern equipment.

It is assumed that a satellite observes all of the plots during the time $\tau_{obs} = 3$ s, so that the time $\tau_{obs1} = 1.5$ µs is spent for viewing one plot. For a superconducting cavity [2], the intensity of oscillations associated with the thermal noise is following:

$$P_{noi} = kT\Delta f_{cav}, \tag{7}$$

where $k = 1.38 * 10^{-23}$ J / degree is the Boltzmann constant; T is the cavity's temperature in Kelvin degrees; $\Delta f_{cav} = 7.5$ kHz is the frequency bandwidth of the cavity. For the temperature $T = 1.4$ K^0 [2], the cavity's noise power is:

$$P_{noi} = kT\Delta f_{cav} = 1.5*10^{-19} \text{ W}. \tag{8}$$

2. Accumulation

When exciting a high-Q cavity by a harmonic signal, the square of the amplitude of its oscillations increases linearly up to values of the order $A^2_{sign} \sim P_{rad} * Q$. During $\tau \approx Q / f_{rad} = 130$ µs the signal grows up to a maximum value. In our case, the time of irradiation and viewing one plot is

1.5 μs, i.e. two orders of magnitude smaller.

For the selected parameters the square of the amplitude of oscillations in the cavity will correspond to the signal power:

$$P_{rad}*Q = 10^{-21}*10^7*1.5*10^{-6}/(130*10^{-6}) = 10^{-16} \text{ W}, \qquad (9)$$

which is almost three orders of magnitude greater than the square of the amplitude corresponding to the cavity's noise power $P_{noi} = 1.5 * 10^{-19}$ W.

Conclusion

From the above it is evident that the coordinates of a moving hot body can be registered reliably both by an infrared telescope and a Doppler radar.

The frequency shift of the reflected signal due to the Doppler effect is proportional to the projection of velocity directed toward or away from the satellite. To avoid dead zones and improve the sensitivity of the method, it is required, obviously, to increase the number of satellites. Such satellites must have different inclinations of their orbits to the Earth's axis.

References

1. http://www.hamamatsu.com/resources/pdf/ssd/infrared_techinfo_e.pdf

2. http://www.linearcollider.org/about/Publications/Reference-Design-Report, http://en.wikipedia.org/wiki/Superconducting_radio_frequency

Registration of the radiation reflected from the water surface at a wavelength $\lambda = 375$ nm

This article considers an opportunity of registering laser radiation diffusely reflected from the surface of the water. It is shown that at a distance of 400 km from the irradiated region it is possible to extract the reflected signal from the background and separate it from the noise. The problem can be solved by means of an array of ten mega pixels consisting of micro pixel avalanche photodiodes operating at room temperature. The article also discusses the background conditions and noise characteristics of the diodes.

Introduction

Let A satellite flying at a height of $H = 300$ km, irradiate the water surface with the light of wavelength $\lambda = 375$ nm. Let the diagram of the emitter be equal to: $\lambda / d_{sat\,A} = 10^{-4}$, so that on the water surface there will be a spot with a diameter $D_{sat\,A} = (\lambda / d_{sat\,A}) * H = 30$ m. We call spot an area of $30 * 30$ m.

Let satellite B fly at the distance $L_{A\,B} = 300$ km from the satellite A at the same height and along the same trajectory.

We consider what set of the equipment must be on the satellite, so that it could record the diffusely reflected radiation from the water surface emitted from the satellite A.

1. The geometric parameters

Let a mirror – lens telescope with an area of a spherical mirror equal to $S_{sat\,2} = 1$ m², be located on satellite B. This corresponds to the diameter of the "plate" equal to: $d_{sat\,B} = 1.13$ m.

Actually, the geometric parameter is the ratio of the area of the mirror telescope located on the satellite to the area of a sphere with a radius $(H^2 + L^2_{AB})^{½}$.

In our case it is:

$$G_1 = S_{sat\,2}/[4\pi *(H^2 + L^2_{A\,B})] = 4.42*10^{-13}. \qquad (1)$$

2. Lambert's law

Lambert's law [1], p. 343, relates the intensity of diffusely scattered light under the angle Θ with the intensity of the reflection under the zero angle of I (0). This law looks as follows:

$$I(\Theta) = I(0) * \cos \Theta. \tag{2}$$

In our case at equal altitude of the satellites and the distance between them equal to their altitude, $\Theta = 45^0$, $\cos \Theta = 0.7$. According to the Lambert law, the radiation intensity at this angle will be 70% of the intensity of the reflected wave at the zero angle.

3. Fresnel formula

The intensity of the reflected wave at normal incidence is equal to: [1], p. 833,

$$I_{ref} = I_{fal} * (n_2 - n_1)^2 / (n_2 + n_1)^2, \tag{3}$$

where $n_2 = 1.35$ which is the breaking index of the water for the wavelength $\lambda = 375$ nm, $n_1 = 1$ – which is the breaking index of the air. Substituting numbers into formula (3), we find that 2.2% of the incident intensity will be reflected from the surface of the water.

Multiplying all three parameters, we find that the overall geometric parameter for our case is equal to:

$$G = 4.42 * 10^{-13} * 0.7 * 2.2 * 10^{-2} = 6.8 * 10^{-15}. \tag{4}$$

4. The number of the registered photons

Let the radiation power of the laser, which is located on the satellite A is $I_A = 1$ kW. In terms of the electron - volts divided by the second, this power will be equal to: $I_{A\,eV} = 6.24 * 10^{21}$ eV / s.

The quantum energy with a wavelength $\lambda = 375$ nm is found from the relation, according to which the quantum of length 100 nm corresponds to the quantum energy of 12.6 eV. Thus, a photon of wavelength $\lambda = 375$ nm will have the energy of $\varepsilon_{375} = 3.36$ eV, and from the satellite A the laser emits the following number of quanta:

$$I_{A\,eV}/\,\varepsilon_{375} = 6.24*10^{21}/3.36 = 1.85*10^{21}. \qquad (5)$$

Taking into account the geometrical parameter, the satellite will take the following number of quanta:

$$N_B = 1.85 * 10^{21} * 6.8 * 10^{-15} = 1.26 * 10^7 \text{ photons / second.} \qquad (6)$$

5. Structure of the registering matrix of the telescope

The satellite A irradiates a spot with a diameter of 30 m on the surface of the water. To register the reflected radiation, it is necessary to view a square under the satellite with the sides each equal to 100 km.

The photo detector array must be a10 Megapixel structure $(3 * 10^3) * (3 * 10^3)$, its square has a side of 1cm. The array consists of columns and rows, wherein each row and each column has $3 * 10^3$ pixels. The size of the side of each pixel in this case is of the order of 3 μ, which is 8 times larger than the limit for the length of wave focusing $\lambda = 0.375$ μ.

The area of the size of 30 * 30 m in this case will be projected to one pixel, and it will sharply differ from the neighboring areas where the intensity of the registered signals from nearby stations will be much less.

When the satellite A irradiates one area, for the satellite B the irradiated area will move at the velocity of 10 km /s. The signal moving over the matrix will run 300 pixels per second, so that each pixel is irradiated with the following number of quanta:

$$1.26 * 10^7/300 = 4.2 * 10^4 \text{ photons / second.} \qquad (7)$$

The photo detector matrix should consist of micro pixel avalanche photodiodes having a single-photo- electron- quantum resolution.

6. Single photo electron quantum resolution

Let us explain why it is so important to have a single photon resolution. Both in the optical and infrared ranges it is impossible to detect a signal from one photo- electron: it is required obligatory to have the signal amplification of the order of 10^6.

Both the useful signal and noises are amplified simultaneously. If the detector has a single photo electron resolution, i.e. the photo electron signal exceeds the amplitude of the noise, and then it is possible to establish such a threshold on the amplifier for to amplify only the signals and cut the noises.

7. The background conditions

In the visible and infrared wavelength range the Sun is a powerful background source. It can be assumed that the Sun, as the irradiator, has a power density 1 kW / m^2 in the wavelength band $\Delta\lambda = 10 \mu$. This power density is more than 3 orders of the power incident from the laser on the analyzed station. Remind that in our case the area is 30 * 30 m; its square is 10^3 m^2.

In order to have the laser radiation power stronger by an order of magnitude in comparison with the background radiation, the laser should have all the power in the wavelength range equal to $\Delta\lambda / \lambda_0 < 10^{-4}$. Extracting the band of the wavelength $\Delta\lambda < 10^{-7} \mu$, is not a big problem for the interferometer [1], p. 225.

8. Noises

Long ago the single photon sensitivity was realized by the company Hamamatsu [3] for the case of the external photoelectric effect. Recently it has performed the single photon resolution for the internal photoelectric effect based on the avalanche photodiode micro pixel. (MLFD) are developed in many research centers, including JINR (Dubna), [4], RF Patents: 1702831, 2102820 and 2086047.

Here is a brief description of the photon detectors of a new generation of micro pixel avalanche photodiodes (MLFD) [4].

APDs allow you to register a new generation of ultra-low light intensities (at the level of single photons). The device was named a micro pixel avalanche photodiode (MLFD). The typical size of each cell (pixel), is about 20 - 30 micrometers, and their number is about 1000 per square millimeter of the working area of the detector. Thus, MLFD as a whole is a device capable of registering the intensity of light (photons) with a dynamic range corresponding to the number of pixels of the photo detector.

Features of the diodes:
- Spectral sensitivity region of 200-900 nm (100 -1200 nm for PMT);
- High quantum efficiency at the maximum of 65-85% (20-35% for PMT);
- Gain of ~ 10^6 (at the level of the PMT);
- Sensitivity at the level of single photons (as for the PMT);
- Photosensitive area of the elements is from 0.5 to 25 mm^2;
- Low operating voltage of diodes - 25-150 V (~ 1000 V for the PMT).

The noises of the micro pixel avalanche photodiodes working at the room temperature can be estimated from the following considerations. The noises (dark current) of the silicon diodes produced by the Hamamatsu Company [3] are 1 pA for a diode with dimensions of 3 * 3 mm. It can be expected that for the crystal with dimensions 3 * 3 μ the dark current will be 6 orders of the magnitude smaller (proportional to the square of the crystal), i.e., it will be of the order of 10^{-18} – or single electrons per second.

Furthermore, the spectrum of the noise electrons lies in the amplitude range smaller than the amplitudes of photoelectrons.

This means that it is possible to register the signals only from the photoelectrons.

Conclusion

Despite the seeming impossibility of registering this scattered radiation, our estimations have shown that it is available. The main difficulty will consist in the accurate determination of the wavelength. Perhaps, the wavelength will be possible to identify quickly and after that – quickly tune the filter cutting the solar radiation.

References

1. Physical Encyclopedic Dictionary, ed. A. M. Prokhorov, Moscow, Soviet Encyclopedia, 1983

2. http://www.meteolab.ru/projects/dielectric/

3. www.hamamatsu.com

4. http://www.jinr.ru

Equipment for detecting the turbulent trace of the body

This article is dedicated to the study of the turbulent trace in the water and air. It is assumed that the turbulent trace of the body in the water is a narrow (100 m) and long (100 km) strip. The trace is supposed to contain self-similarity elements and, accordingly, has a fractal component. To determine this fractal component, the water surface is necessary to irradiate with the laser and register the reflected wave. The article gives estimates of the required radiation intensity and sensitivity of the detectors.

The required sensitivity of the equipment to detect the turbulent trace in the air has been estimated. It is shown that the trace at a height $h = 50$ km with a diameter $d_{wake} = 10$m will be visible at a distance of 1000 km.

I. Turbulent trace in the water

We consider possible results of fractal analysis of the series of numbers obtained by digitizing of the signal from the interferometer. To one arm of the interferometer the first signal arrives as a result of reflection of the electromagnetic wave from the surface of the water. The reference signal arrives to the other arm of the interferometer. The emitter of electromagnetic waves having a narrow radiation pattern consistently exposes different parts of the surface of the water. The receiving equipment detects the signal reflected from each part of the water surface.

The reflected signal is summed up with the reference signal, and in the result the total intensity - the amplitude of the resulting oscillation, has the following dependency on the amplitudes and phases of the summed up signals:

$$A^2 = a_1^2 + a_2^2 + 2a_1a_2\cos(\varphi_1 - \varphi_2), \qquad (1)$$

where A - amplitude of the resulting oscillations, a_1 - the amplitude of the reference wave, a_2 - the amplitude of the measured wave, φ_1 - the phase of the reference wave, φ_2 - phase of the measured wave. Thus, from (1) it follows that the square of the amplitude of the resulting oscillation is proportional to the intensity of the detected signal depends on the phase difference $(\varphi_1 - \varphi_2)$, on the reference and studied signals.

Digitizing this intensity, we get a series of numbers containing the information about the phase φ_2, of the oscillation being studied.

41

1. Selection of the basic parameters

We choose the height of the transmitter (and the receiver) above the water surface equal to H = 10 km. Let the irradiated portion have a size of 100*100m.

We choose the diameter of the receiving plate equal to d_{pl} = 0.84 m and the radius of its bending equal to r_0 = 8.4 m, so that the focal length of the spherical mirror will be equal to F = r_0 / 2 = 4.2 m. The geometric parameter for this spherical mirror is: $(\pi d_{pl}^2 / 4) / 4\pi H^2 = 5 * 10^{-10}$. Assume that for an area of 100 * 100 m obtains power W_{fall} = 1 kW.

According to Fresnel formulae for normal incidence, we have [1], p. 883,

$$r_{refl} = (n_2 - n_1)^2/(n_2 + n_1)^2, \tag{2}$$

where n_1 is the breaking coefficient of the first medium, n_2 - the breaking coefficient of the second medium. In this case, n_1 = 1 - air, n_2 = 1.3 - water, [2].

Substituting numbers into the formula (2), we find that due to the difference between the breaking coefficients of air and water, in the case of normal incidence only 1.7% of the power incident from the air onto the water is reflected from the water.

We assume that the diffusely reflected power is 1% of the incident power, then taking into account the geometrical parameter of $5 * 10^{-10}$, the received power is as follows:

$$W_{res} = W_{ir}*5*10^{-10}*10^{-2} = 5*10^{-9} W, \tag{3}$$

where W_{ir} = 1 kW is the power radiated by the laser and the incident onto the area of the water surface with a size of 100*100 m, W_{res} is the received power.

The sensitivity of the modern InGaAs infrared detectors [3] is of the order: $D* = 5 * 10^{12}$ (cm * $Hz^{1/2}$)/ W. The value inverse to $D*$ is the minimum level that can be registered by this detector.

In our case, the value 1 / $D*$ is 1 / $D* = 2 * 10^{-13}$ W / (cm * $Hz^{1/2}$). For the detector having an area of 10^{-4} cm^2 and a band Δf = 300 Hz, the minimum sensitivity of the detector will be $P_{min} = 3 * 10^{-14}$ W.

Let us explain why we have taken $\Delta f = 3 * 10^2$ Hz. Totally we intend to study the area equal to 10 * 0.3 km. This area contains 300 areas with dimensions of 100 * 100 m, and it is needed to observe them during one second. Thus, the data taking rate should be 300 areas per second, from where we have obtained $\Delta f = 3 * 10^2$ Hz.

2. Interferometer

In the visible and infrared wavelength range a powerful background source is the Sun. It can be assumed that the Sun, as a radiation source, has the power density equal to 1 kW / m² in the wavelength band $\Delta\lambda = 10$ μ. This power density is by 4 orders of magnitude more than the power incident from the laser to the area under study.

In order to have the laser radiation power by the order higher than the background radiation, the laser should emit the full power in the wavelength range $\Delta\lambda / \lambda_0 < 10^{-4}$. The irradiated power must be equal to $W_{irr} = 1$ kW, that is a strong requirement to the parameters of the laser.

The loss of intensity of the wave under study during the passage of the interferometer to the area of summing up the signals is assumed to be 10^{-1}. The allocation of the wavelength area $\Delta\lambda < 10^{-7}$ μ does not seem to be a big problem for the interferometer [1], p. 225.

It should be keep in mind that the dynamic range of the signal under study should be at least 10^3, this method gives the sensitivity of the apparatus to be quite sufficient.

3. The structure of the observed signals

Thus, if the minimum detector sensitivity $P_{min} = 3 * 10^{-14}$ W the signal before the interferometer will be $5 * 10^{-9}$ W. After passing through the interferometer just before summing up the signals, the power of one signal is equal to $5 * 10^{-10}$ W and this is approximately by 4 orders greater than the minimum sensitivity of the detector.

At the observation time for one area equal to 3 ms and ADC speed equal to 1 μs, each area will send a numerical row, consisting of approximately $3 * 10^3$ figures. The obtained values express the intensity of the summed up photo current with a sufficiently high accuracy.

4. Ways to improve the sensitivity

Transition of the temperature of the detectors from $T_{det} = 300$ K^0 to the liquid-helium temperature $T_{det1} = 3$ K^0 will reduce the power of the thermal noise $kT\Delta f$ by about 100 times. It is necessary to say that besides thermal noises there are other types of noise. It is important that, at least, the thermal noise of the detector will reduce by100 times.

The wave limit of focusing the radiation with a wavelength of $\lambda_0 = 1.6$ μ is equal to the wavelength, i.e, possible reduction of the size of a pixel is approximately 60 times.

II. Turbulent trace of the body in the air

Suppose that at a height $h = 50$ km there is a moving body which leaves a turbulent trace behind itself. The pressure at height $h = 50$ km is approximately $P_{atm1} = 0.64$ mm. Hg, that is approximately 10^3 times smaller than the normal atmospheric pressure of 760 mm. Hg.

If the initial size of the trace had a diameter of 30 cm, and the pressure therein was about normal atmospheric pressure, then at the expansion of the trace the pressure in it will drop. When the trace diameter reaches 10 m the pressure becomes equal to the pressure in the surrounding air, and further trace expansion will slow down. Further the trace will continue to expand due to the extra partial pressure in it and also due to diffusion.

The range of direct vision of the body located at height $h = 50$ km from the surface of the Earth is $l_{obs1} = (2\ h\ R_E)^{1/2} \approx 800$ km. If to lift the observer to a height $h_1 = 12.5$ km, for him the direct vision range will be equal to 400 km and the body located at an altitude of $h_2 = 50$ km can be seen by the observer at a distance of $l_{obs2} = 1200$ km.

We assume that the distance between the observer and the body, leaving the turbulent trace is $l_{obs3} = 1000$ km.

Let the observer record the laser radiation reflected from the turbulent trace of the body. Assume the laser direction be equal to $\Theta_{las} = 10^{-2}$. Then, at the distance of 1000 km from the observer the radiation will occupy an area with dimensions of 10*10 km. Let the observer look through at a strip above the horizon, with a height of 10 km and a width of 1000 km, then the band has 10^3 areas with dimensions of 10*10 km.

The turbulent trace of the body on one of these areas covers a square of $S_{wake} / S_{plot} = 10$ m * 10 km / 10 * 10 km $= 10^{-3}$ of the total area. We assume that the vertical size of the trace is 10 m, and the horizontal length is equal to the length of the area equal to 10 km.

Let the power of the laser irradiating consequentially all the areas be $P_{las} = 1$ kW. Let all the 10^3 areas are irradiated by the laser during 1 s. Then to irradiate one area, it is necessary to spend $\Delta\tau_{plot} = 1$ ms, and this area is irradiated with the energy of 1 J.

Let the wavelength of the laser radiation is $\lambda_{las} = 0.5$ μ. The energy of one photon with this wavelength is equal to $\varepsilon_{las} = 2.5$ eV, and thus, each area is irradiated with 1J / $\varepsilon_{las} = 2 * 10^{18}$ photons.

Let photons are reflected isotropically from each area. Then the geometric parameter for the receiving telescope with a mirror diameter $D_{tel} = 1.3$ m is equal to $\pi D^2_{tel} / (4 * 4 * \pi * 1^2_{obs3}) = 10^{-13}$.

The total geometric and temporal parameter of reducing the received power in comparison with the power of radiation will be equal to $10^{-3} * 10^{-13} = 10^{-16}$.

This means that each area will give to the array of the receiving telescope about $2 * 10^{18} * 10^{-16} = 10^2$ of photons.

The single photon sensitivity was realized by the same company Hamamatsu [3] long time ago for the case of the external photoelectric effect. Recently, for the case of the internal photoelectric effect they obtained the single photon sensitivity for avalanche photodiode micro pixels. APDs were developed in many research centers, including the JINR [4], RF Patents: 1702831, 2102820, and 2086047.

APDs are photodiodes of a new generation which allow one to register ultra-low light intensities (at the level of single photons). The device was named a micro pixel avalanche photodiode (MLFD). The typical size of each cell (pixel) is about 20 - 30 micrometers, and their quantity in the diode is of the order 1,000 per square millimeter of the working area of the detector. Thus MLFD is a device capable to register the intensity of light (photons) with a dynamic range corresponding to the number of pixels of the photo detector.

Features:
- Spectral sensitivity region of 200-900 nm (100 -1200 nm PMT);
- High quantum efficiency at the maximum of 65-85% (20-35% for PMT);
- Gain of $\sim 10^6$ (at the level of the PMT);
- Sensitivity at the level of single photons (at the level of the PMT);
- Photosensitive area elements from 0.5 to 25 mm^2;
- Low operating voltage of 25-150 V ($\sim$ 1000 V for the PMT).

The noises of micro pixel avalanche photodiodes working at room temperature can be estimated from the following considerations. The noises (dark current) of silicon diodes produced by company Hamamatsu [3] are 1 pA for the diode with dimensions of 3 * 3 mm. It can be expected that for the crystal with dimensions 3 * 3 μ it will be by 6 orders of magnitude smaller (proportional to the area of the crystal), i.e. it will be the value of the order of 10^{-18} A, it is the number of single electrons per second. Moreover, the spectrum of the noise electrons lies in the amplitude range smaller than the amplitude of one, two and so on photoelectrons.

This means that it is possible to establish the detection threshold greater than the noise signals and register the signals only from photoelectrons.
To suppress the background of the solar radiation, it is necessary to use a narrow-band filter with a bandwidth equal to the bandwidth of the laser radiation. We can use a Fabre – Perot interferometer as this filter.

The loss of intensity of the wave under study during its passing through the interferometer is assumed to be equal to 10^{-1}. Selection of the band of wavelengths $\Delta\lambda < 10^{-7} \mu$, is not a big problem for the interferometer [1], p. 225. Then in the narrow band of wavelengths reflected from the turbulent trace the laser ray will significantly exceed the brightness of the scattered solar radiation.

The number of pixels in the vertical rows of photo detector array must allow the matrix to extract the thin trace (10 m) at the height of the area equal to 10 km. This means that the number of pixels in a vertical row must be greater than 10^3.

Conclusion

From the above estimates it follows that by means of the mirror - lens telescope with a photo detector array consisting of micro pixel avalanche photodiodes it is possible to find the turbulent trace 10 m thick at a distance of 1000 km.

References

1. Physical Encyclopedic Dictionary, ed. A. M. Prokhorov, Moscow, Soviet Encyclopedia, 1983

2. http://www.meteolab.ru/projects/dielectric/

3. www.hamamatsu.com

4. http://www.jinr.ru

Technique

A directed underwater explosion

To form an explosion wave with a flat or concave spherical front, the explosive mixture is presented as a spiral with a constant or variable pitch and radius of the winding. Then the velocity of propagation of the explosion wave becomes smaller than the propagation velocity of the detonation wave. The explosion wave velocity can be varied by changing a pitch value or the winding radius of the spiral. These spirals are located over the lateral surface of the cone. If the winding pitch of the spiral $h_{spiral} = 40$ cm, the winding radius $r_0 = 25$ cm, the diameter of the hexogen thread $d = 1$ cm, the number of spirals is $n = 36$ spirals on the lateral surface of a cone with a vertex angle $\Theta = 45^0$ and base diameter of the cone $D_{cone} = 150$ m, the mass of the explosive mixture is of the order of 2 tons. At a distance $r = 20$ km from the cone the pressure on the front of the explosion wave is $P > 2$ atm.

Introduction

The underwater explosive wave is sharp compression of water.
The beginning of the explosive border wave is called the front. Here the pressure has the maximum value.

At the moment when the front of the underwater explosive wave comes to a given point the water pressure in this point increases from the hydrostatic value to the maximum. The object placed in this point obtains a sharp blow.

Below we will consider the conditions of forming the explosive wave with a flat or concave spherical front. Let us dispose in the explosive mixture in the form of a spiral with a constant or variable pitch and some radius of the spiral winding. The propagation velocity of the explosive wave along the spiral axis becomes smaller than the propagation velocity of the detonation wave. This propagation velocity of the explosive wave along the spiral axis can be changed by both – varying the pitch or the radius of the spiral winding.

Now we arrange these spirals gradually over the lateral surface of the cone. The angle at the vertex should be chosen for the projection of the velocity of the explosive wave on to the height of the cone, to be equal to the sonic velocity in the water. Such a sonic wave will come to the focus point during the same period of time from the explosion of different points of the spirals.

1. Scheme of spiral locations

FIG. 1 schematically shows the relative arrangement of the spirals.

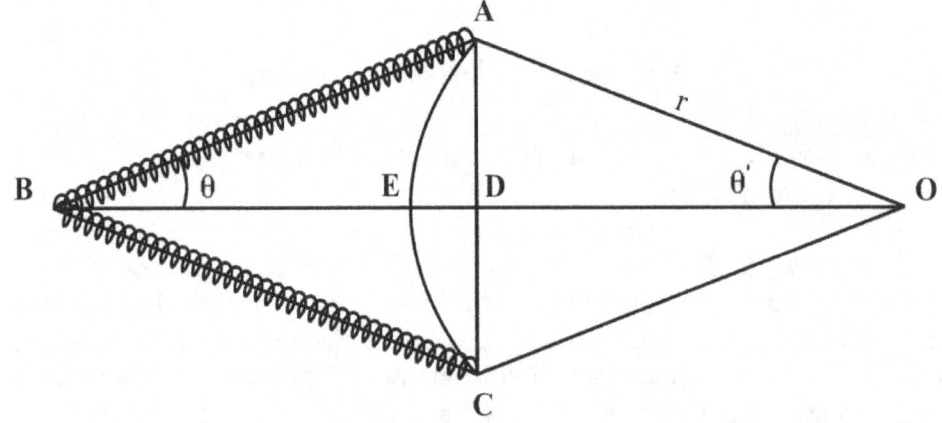

Fig. 1 Scheme of spiral locations

Triangle ABC shows a longitudinal section of the cone with angle Θ at the vertex B. The lateral surface of the cone (the drawing segments BA and BC) is laid with hexogen threads rolled into a spiral. The total number of these spirals is n = 36. They are arranged azimuthally at regular angular intervals, in this case $\Delta\Theta = 10^0$. To obtain the ratio of the projection of the velocity of the explosive wave propagation on to the height of the cone equal to the velocity of sonic propagation in the water, we have to select properly the cone vertex angle

48

Θ, the pitch and radius of the spiral winding.

Then the sonic wave from the explosion of any point of the spiral comes simultaneously to all the base of the cone, (Fig. points A, D and C). If to wind the spiral with a variable pitch and radius of the winding, it is possible to form a spherical concave front of the explosive wave, whose cross section is shown in Figure by points A, E, C. This spherically converging sonic (explosive) wave will focus at the point located at radius r from the cone. When the convergence angle Θ 'is much smaller than Θ, the focusing radius will be much larger than the height of the cone.

2. Selection of the basic parameters

2. 1. Dynamics of explosion

For the water the compressibility coefficient is [1], p. 71 $k_{water} = 5*10^{-5}$ atm^{-1}. For the pressure lower than 10^3 atm., the water density in the explosive wave slightly differs from the water density under normal conditions and the velocity of the explosive wave is approximately equal to the sonic wave velocity:
$V_{water} \approx V_{sound} = 1.5$ km / s [1] p. 83.

2. 2. Selection of the cone vertex angle

2. 2. 1. A solid hexogen cone

We consider conditions of forming the directed explosive wave by means of the solid hexogen cone. Let us find cone vertex angle.

The velocity of the detonation wave in RDX, $V_{det.\ wave} = 8.36$ km / s [2], greatly exceeds the velocity of the sound in the water: $V_{sonic} = 1.5$ km / s. To form a flat front of the sonic wave, it is necessary to choose the cone angle equal to the following:

$$\Theta_{plane\ front1} = \arccos V_{sonic}/ V_{det.\ wave} = 0.18. \qquad (1)$$

The corresponding angle between the velocity of the detonation wave and the sonic velocity in this case is equal to: $\Theta_{plane\ front\ 1} = 80^0$.

Let the diameter of the cone base be $D_{cone} = 150$ m, its wall thickness $\Delta h_{cone} = 1$ cm. Assuming that the length of the side of the cone is approximately equal to half the diameter of its base, we find the square of the lateral surface of the cone to be approximately equal to the base square:

$S_{lat.\ surf} = \pi D_{cone}^2 / 4 = 7 * 10^4$ m^2. The volume occupied by hexogen is $V_{cone} = 700$ m^3. Assuming the density of RDX [2] to be equal to: $\rho_{hex} = 1.8$ g / cm^3, we find that its total mass is equal to the following: $M_{hex} = \rho_{hex} * V_{cone} = 1260$ tons.

While undermining this cone from the vertex to the plane of its base, the flat front of the explosive wave is formed. However, the use of such a large number of explosives seems to be excessive and the front duration will be very short: $\tau_{exp.} = \Delta h_{cone} / V_{det.\ wave} \approx 1$ μs. This will result very soon in rapid dissipation of energy of the sonic wave due to attenuation [3]. As it is shown in [3], the range of the sound spreading out with frequency $f_{sonic} = 4 / \tau_{blast} = 0.25$ MHz is less than 1 kilometer.

The sound spreading out with frequency $f_{sonic} = 2.5$ kHz, reduces its intensity by 10 times at a distance of 100 km [3]. To increase the duration of the explosion front by 2 orders, it would have been necessary to increase the wall thickness of Δh_{cone} cone also by 2 orders. It would have been resulted in increasing the mass of the explosive that is not desirable at all.

2. 2 .2. Cone laid out with spirals

We choose the cone angle equal to: $\Theta_{plane\ front\ 2} = 45^0$. To get this angle between the direction of the explosive wave and the velocity of the sound in water, it is necessary to slow down the propagation velocity of the explosive wave.

The slow down process of the explosive wave in the spiral has a purely geometric nature. The detonation wave propagates over the spiral with the velocity $V_{det.\ wave} = 8.36$ km / s. The propagation velocity of the explosive wave along the axis of the spiral is smaller and equal to the following:

$$V_{exp.\ wave} = (h_{spiral}/2\pi r_0) * V_{det.wave}, \qquad (2)$$

where h_{spiral} – the spiral winding pitch, r_0 - the radius of the spiral.

We choose the radius of the spiral winding equal to: $r_0 = 25$ cm. Then the diameter of the hexogen spiral is $d_{spiral} = 50$ cm. To have the angle between the velocity of the explosive wave and the velocity of the sonic wave to be equal to:

$\Theta_{\text{plane front 2}} = 45^0$, it is necessary to have the explosive wave propagation velocity to be equal to $V_{\text{exp. wave}} = 2.14$ km / s. The spiral pitch should be equal to:

$$h_{\text{spiral.}} = (2\pi r_0 / V_{\text{det.wave}}) * V_{\text{exp. wave}} = 40 \text{ cm}. \qquad (3)$$

Let the diameter of each hexogen thread be equal to $d_{\text{thread}} = 1$ cm. The length composing the cone exceeds the half diameter length of the cone by 1.5 and is equal to:

$$l_{\text{gen}} = D_{\text{cone}} / 2 * \cos 45^0 = 107 \text{ m}. \qquad (4)$$

Along the cone composing it is possible to allocate the following number of turns of the spiral: $n = l_{\text{gen}} / h_{\text{spiral}} = 267$. The total length of the spiral is equal to $l_{\text{spiral}} = 420$ m, the length of one turn is $2\pi r_0 = 1.57$ m.

The volume of each spiral will be equal to:
$V_{\text{spiral}} = (\pi d_{\text{thread}}^2 / 4) * l_{\text{sp}} = 3.3 * 10^{-2}$ m^3.

The mass of one hexogen thread will be equal to:
$M_{\text{thread}} = \rho_{\text{hex}} * V_{\text{spiral}} = 60$ kg. The total mass of RDX in $n = 36$ coils located on the lateral surface of the cone is equal to: $M_{\text{hex}} = M_{\text{thread}} * 36 = 2.16$ tons. The hexogen energy released when the RDX explosion, is equal to [2], 1.37 kcal / kg or 5.75 GJ / ton. Such a quantity of the explosive located on the lateral surface of the cone, seems to be acceptable.

3. Spatial and temporal duration of the explosive wave front

The temporal and spatial duration of the front of this explosive wave will be short, of the order of $\tau_{\text{front}} = \pi r_0 / V_{\text{det.wave}} \approx 0.1$ ms. This is related with the fact that different parts of the spiral explode sequentially, and the projection of the velocity of the explosive wave to the height of the triangle is exactly equal to the sonic velocity. From two points located on the diameter, at the same distance from the focus, the explosive wave propagating along the height of the triangle, will come with the following delay: $\tau_{\text{front}} = \pi r_0 / V_{\text{det.wave}} \approx 0.1$ ms.

Considering the magnitude of τ_{front} half period to correspond to this duration of the sonic wave, we find that the sonic wave period T_0 is:
$T_0 = 4 * \tau_{\text{front}} = 0.4$ ms. The frequency of the sonic wave is then equal to: $f_{\text{sonic}} = 2.5$ kHz. The distance at which the sound wave of this frequency decreases its intensity by 10 times, is of the order of 100 km [3].

51

This sonic wave frequency $f_{sonic} = 2.5$ kHz corresponds to the following wavelength: $\lambda_{sonic} = V_{sonic} / f_{sonic} = 0.6$ m.

4. Propagation of the flat front of the sonic wave from the cone with a vertex angle equal to $\Theta_{plane\ front\ 2} = 45^0$

Consider the propagation of a sonic wave produced by the flat front. Its diffraction limit, i.e. and its minimum divergence angle of the flay front will be equal to:

$$\Theta_{exp.} = (\lambda_{sonic}/D_{cone}). \tag{5}$$

The size of the area occupied by the sonic wave will be equal to:

$$\Delta y = \Theta_{exp.}*F = (\lambda_{sonic}/D_{cone})*F, \tag{6}$$

where F- is the distance from the cone to the observation point, in our case $F = 20$ km.

Substituting numbers into the formula (6): $\lambda_{sonic} = 0.6$ m, $D_{cone} = 150$ m, $(\lambda_{sonic} / D_{cone}) = 4 * 10^{-3}$, we get $\Delta u = \Theta_{exp.} * F = (\lambda_{sonic} / D_{cone}) * F = 80$ m.

5. The pressure at the front of the sonic wave

We estimate the pressure at the front of the sonic wave from the following considerations. The total released energy of the explosion is equal to:

$$\Delta Q = M_{hex}* 5.75 = 12 \text{ GJ.} \tag{7}$$

At the pulse duration equal to: $\tau_{front} = 0.1$ ms the power of the explosion will be equal to:

$$W= (\tfrac{1}{2})\Delta Q/ \tau_{front} = 60 \text{ TW.} \tag{8}$$

Thus, the intensity of the front sonic wave can be estimated as follows:

$$I = W/S = \Delta Q/ 4\pi F^2*\tau_{front} = 2.5*10^4 \text{ W/m}^2. \tag{9}$$

The pressure in the sonic wave is equal to:

$$P = (I_{front} * \rho_{water} * V_{sonic\ wave})^{1/2} \approx 2*10^5 \text{ Pascal} = 2 \text{ atm.} \qquad (10)$$

Such a pressure at distance F = 20 km from the explosion point, is of practical interest for some applications.

6. Formation of the concave spherical front of the sonic wave from the cone with a vertex angle equal to $\Theta_{plane\ front\ 2} = 45^0$

If by means of the point explosion it is possible to obtain only a convex spherical front, in our case we can receive a concave spherical front.

To obtain a flat front, it is necessary to have a constant propagation velocity of the explosive wave. To obtain a concave spherical front, it will be needed to form a variable velocity which increases while spreading out of the explosion wave.

We find conditions under which the time of arrival of the sonic wave front into the focal point O will be the same at the explosion of any point on the spiral, i.e, of any point on the interval AB or BC on the lateral surface of the cone.

We denote the length of BO - from the vertex to the focus of the spherical front to be the size of l_0. The distance AB has already been named the value l_{gen}, then:

$$l_0 = l_{gen} * \cos\Theta + (r^2 - l^2_{gen} * \sin^2\Theta)^{1/2}. \qquad (11)$$

Let us denote the time during which the sonic wave travels the distance l_0 as t_0, so that

$$t_0 = l_0 / V_{sonic} \qquad (12)$$

During the same time t_0 the sonic wave should reach point O from any point on the spiral. The distance from point B to this certain point is called x. Then the distance between the point x and the point O (the center of the sphere) is equal to r_1:

$$r_1^2 = (x * \sin\Theta)^2 + (l_0 - x * \cos\Theta)^2. \qquad (13)$$

When you remove the focal point O at a distance much greater than the height or the lateral side of the cone, the concave spherical wave front is almost

53

flat. It means that the explosion wave propagation velocity V (x) along the lateral side of the cone will be almost equal to V- explosion wave velocity. Assume $V(x) = V_{exp.\,wave} + \delta V$, where $\delta V \ll V_{exp.\,wave}$.

Let us find δV from the following relation:

$$V(x)*\cos\Theta = V_{exp.\,wave}*\cos\Theta + \delta V*\cos\Theta. \qquad (14)$$

For the period of time corresponding to the passing of the explosive wave along the lateral cone side $t_{lat.\,surf} = l_{gen} / V_{exp.\,wave}$, the difference between $V(x) * \cos\Theta * t_{lat}$ and $V_{exp.\,wave} * \cos\Theta$, which is equal to $\delta V * \cos\Theta * t_{lat}$ is equal to **a** - the height of the chord:

$$a = [r^2 - (r^2 - l^2_{gen}*\sin^2\Theta)]^{1/2}. \qquad (15)$$

From the above:

$$\delta V = [r^2 - (r^2 - l^2_{gen}*\sin^2\Theta)]^{1/2}/\cos\Theta * t_{lat}. \qquad (16)$$

The dependence of the explosive wave propagation velocity $V_x(x)$ is weak and can be selected, for example, as $\delta V_1 = \delta V * x / l_{gen}$, so that when $x = l_{gen}$ which is an addition to velocity δV_1 equal to δV.

If the velocity of the explosive wave in $V_{exp.\,wave}(x)$ changes according to the law during propagation along the lateral surface of the cone:
$V(x) = V_{exp.\,wave} + \delta V$, $\delta V_1 = \delta V * x / l_{gen}$,
$\delta V = [r^2 - (r^2 - l^2_{gen} * \sin^2\Theta)]^{1/2} / \cos\Theta * t_{lat}$, then the arrival of the sonic wave (the explosive wave front), to point O is the same for all points on the lateral surface of the cone.

To change the velocity of propagation of the explosive wave along the conical surface, can be performed by changing the pitch or/and the radius of the spiral winding according to formula (2).

7. Focal length

If the sonic wave propagation velocity along the height of the cone with an angle at the vertex equal to: $\Theta_{plane\,front\,2} = 45^0$, is less than the velocity of the explosive wave propagating along the axis of the spiral on the lateral surface of the cone, then the concave front of the sonic wave is formed in the base of the cone.

If to make a transverse cross-section of the cone, we get a triangle. Let us denote the points A and C as the points lying at the base of the triangle and the point B lies in its vertex. Let point D lie at the intersection of the base of the triangle with its height. Then, the point E, which lies on the same circle of radius r as points A and C, is inside the triangle ABC. We denote the distance ED as the value **a** – it is the difference between the flat and concave fronts on the axis. Let us denote the distance AD as value **b**: it is the half of the width of the cone base. Point O denotes the center of the circle of radius r, passing through AED. Angle AOE is denoted with value Θ'.

Then: $(r_a) / r = \sin \Theta'$, $b / r = \cos \Theta'$, we write the following identity: $\sin^2 \Theta' + \cos^2 \Theta' = 1$, in the form of:

$$(r-a)^2/r^2 + b^2/r^2 = 1, \tag{17}$$

from where we find the radius of the circle:

$$r = (a^2 + b^2)/2a, \tag{18}$$

and taking into account that $a << b$, we get:

$$r \approx b^2/2a. \tag{19}$$

So, if we want to focus the explosive wave at a distance $r = 20$ km, for $b = D_{cone} / 2 = 75$ m, we will have to preserve the precision of the following parameters: $a = b^2 / r \approx 0.3$ m.

The pressure in the focus of such convergent sonic wave is, at least, not less than the pressure at the front of the flat sonic wave: $P > 2$ atm.

The period of time during which the explosive wave is propagating at a distance of $r = 20$ km, will be a little bit more than 13 seconds.

References

1. Tables of physical quantities. Handbook ed. I. K. Kikoin, Moscow, Atomizdat, 1976

2. http://ru.wikipedia.org/wiki/Гексоген

3. http://www.akin.ru/spravka/s_ocean.htm

The flight trajectory control of the body

The article is dedicated to the control of the flight trajectories of the bodies moving with space velocities. This proposal can be used to correct the trajectory of rockets if the contact with them or their targeting self- systems has been lost.

Onto the lateral surface of the body with mass of $m_{body} = 1$ ton moving at $V_{body} = 10$ km / s, we paint four bands of silicon with different degrees of doping. The square of each strip is $S_{strip} = 1 m^2$, the mass of $m_{strip} = 1$ kg. Absorption of laser radiation leads to the evaporation of the strip and heating of its atoms up to their thermal velocity $V_{heat} = 2.5$ km / s. The average jet velocity in this case is of the order of $V_{jet} = 1$ km / s, that results in the body deviating by the angle $\theta^{\perp} = (p^{\perp} / m_{body} * V_{body})$ from the initial trajectory.

Introduction

To control the body, on its lateral surface we paint four strips of the material having different frequencies of the resonant absorption of laser radiation. The control is carried out by means of deviation of the body to the "right-left" and "up - down". The deviation is obtained by evaporation of the corresponding band while its exposure to the laser radiation with a resonant frequency.

1. Physical motivation of the task

To control the flight trajectory of the body, its side surface is painted with four bands, for example, of silicon having a different doping level. Then their Langmuir frequencies ω_{pl} will differ and, accordingly, these four bands will have a different frequency of resonant absorption [1].

Thus, at the resonance absorption of the laser radiation only one of four bands evaporates and gives a reactive jet. The body deviates namely in the opposite direction from this jet.

In order to evaporate the strip for a short period of time, it is necessary to have the laser impulse to be rather short.

Fig. 1 schematically shows the relative arrangement of elements.

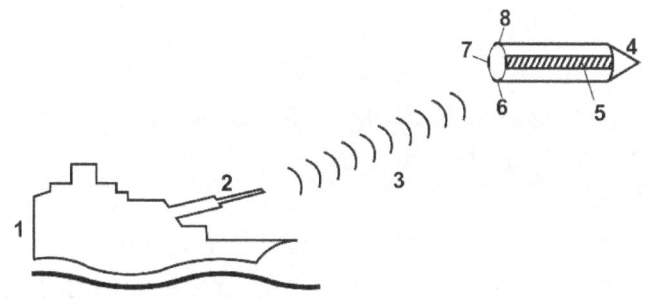

Fig.1. Location of the elements in the space.

On a floating platform (1) there is laser (2), whose radiation (3) is incident onto the body (4), and one of the strips (5), (6), (7), (8), deposited on a lateral side surface of the body resonantly absorbs this radiation and evaporates. This gives a reactive jet, while for the other bands the reflectance coefficient of laser radiation is close to 1, and these strips do not interact with laser radiation.

2. Selection of the basic parameters

2. 1. The flight parameters of the body and its trajectory changes

Let the body has a mass of $m_{body} = 1$ ton and is moving at a velocity of about $V_{body} = 10$ km / s. At the expiring of the jet mass equal to 1 kg. At the velocity $V = 1$ km/s, the transverse momentum transmitted to the body will be equal to: $p^{\perp} = m_{jet} * V_{jet}$. This will lead to the appearance of the deviation angle of the body equal to: $\theta^{\perp} = p^{\perp} / (m_{body} * V_{body}) = 10^{-4}$. Such an angle at a distance from the target: $s_1 = 1000$ km, will result in deviation of the body by the distance of 100m from the unperturbed trajectory of the body: $\Delta l = s_1 * \theta^{\perp} = 100$ m.

2. 2. Energy relations for the jet expiring

The heat capacity of silicon is, $c_{Si} = 20$ J / (mol * degree) [2], p. 199, the melting point $T_{melt} = 1415$ ^{0}C, the heat of phase transition from solid to liquid $\Delta W_{melt} = 50$ kJ / mol, the evaporation temperature: $T_{eva} = 3300$ ^{0}C, the heat of the phase transition from the liquid to vapor: $\Delta W_{eva} = 356$ kJ / mol [2], p. 289. Summing up all the energy necessary for the evaporation of silicon and taking into account that 1 mole of silicon is 28 g, we find that for the evaporation of 1 gram of silicon it is required to expend the energy of ~ 15 kJ / g.

For the average directed velocity of the silicon atoms would be equal to: $V_{jet} = 1$ km / s, it is necessary to have the thermal velocity to be equal to:

$V_{heat} = 2.5$ km / s. Indeed, after averaging the velocity in one of the transverse planes, we obtain the following: $V^{\sim}_1 = (V_T / \pi) * \int \sin\varphi \, d\varphi = (2 / \pi) * V_{heat}$, where the integration over the angles must be carried out from 0 to π. After averaging the velocity in the two transverse planes we get:
$V^{\sim}_2 = V_{jet} = (2 / \pi)^2 * V_T \approx 0.4 \, V_{heat}$. Thus, besides the evaporation of silicon it is necessary to transfer thermal velocities of the atoms $V_{heat} = 2.5$ km / s to silicon in order to have the jet average directed velocity to be equal to
$V_{jet} = 1$ km / s.

We find the energy of the silicon atom, moving with a velocity of V_{heat} from the ratio: $m_{Si} * V_{heat}^2 / 2 = \varepsilon_{Si} = 1.5 * 10^{-19}$ J. Taking intro account that 1 g contains $2 * 10^{22}$ atoms, we find that it is required additionally to invest the energy of the order of 3 kJ / g, and the total energy consumption should be equal to the following: $W_{las} \sim 20$ kJ / g.

2. 3. *Parameters of the body exposure*

We consider an opportunity of the pulse CO_2 laser at a distance $s_2 = 1000$ km from the body being exposed. For the diffraction divergence of the laser ray would be small enough, it is necessary that the individual laser emitters have been combined into a laser lattice [3]. It is similar to the way how the individual radio emitters are synchronized by the phase in the phased array antenna lattice operating at radio frequencies.

Let the total diameter of the laser lattice be equal to: $d_{las} = 3$ m. Then the angle of the diffraction divergence will be equal to: $\theta_{dif} = \lambda / d_{las} = 3 * 10^{-6}$, where $\lambda = 10 \, \mu$ is the wavelength of the laser radiation. Thus, at a distance $s_2 = 1000$ km the square of the laser spot can be estimated as $S = \pi * (s_2 * \theta_{dif})^2 = 30$ m^2.

Let the square of a single band on the body, which is to be evaporated to reject the body by an angle $\theta^{\perp} = p^{\perp} / (m_{body} * V_{body}) = 10^{-4}$, be equal to: $s_{strip} = 1$ m^2. Then the geometric parameter in this case is equal to: 1/30. So, for heating and evaporating of one gram of silicon it is required to have the laser energy $W_{las1} = 20$ kJ. Taking into account the geometrical parameter the laser energy should be by 30 times more and equal to $W_{las2} = 600$ kJ.
To heat and evaporate 1 kg of silicon, it is required to have the laser energy by 1000 times greater: $W_{las3} = 600$ MJ.

Conclusion

Thus, irradiating the body with the laser located at a distance from the body equal to $s_2 = 1000$ km, it is possible to transfer a transverse momentum to the body by evaporation of one of four bands, deposited on the body: $p^{\perp} = m_{jet} * V_{jet}$. It will result in the appearance of the deviation angle: $\theta^{\perp} = p^{\perp} / (m_{body} * V_{body}) = 10^{-4}$. This angle at the distance from the target: $s_1 = 1000$ km, will lead to deviation of the body from its unperturbed trajectory at a distance: $\Delta l = s_1 * \theta^{\perp} = 100$ m.

References

1. V. Y. Timoshenko, Optics nanosystems, Lecture 4, Exiton and impurity absorption of light, Moscow State University, Research and Education Center for Nanotechnology, http://nano.msu.ru/files/systems/V/autumn2011/optics/Timoshenko_L04_NOC 2011 .pdf

2. Physical quantities Handbook, ed. I. S. Grigoriev and E. Z. Meilikhov, Moscow, Energoatomizdat, 1991

3 V.P. Kandidov, Laser arrays, http://www.pereplet.ru/nauka/Soros/pdf/9912_068.pdf

The tethered balloon with a useful load

The article considers the conditions under which it is possible to place a captive balloon, located at the altitude $H_a = 7$ km with a useful mass $m_{load} = 1$ ton. This load is supplied with power $P \approx 2$ kW. It is shown that at the wind velocity of 4 m / s the lateral wind load will lead to shifting of the balloon at a distance of 10 m.

Introduction

The durability of the conventional steel is of the order of 40 kgf/mm^2 or 4 ton/cm^2 [1], p. 38. Considering the density of iron to be equal to $\rho_{Fe} = 8$ g/cm^3, we find that the one-meter iron rod with a cross section of 1 cm^2 weighs 800 g, one kilometer of the rod weighs 800 kg and the rod 5 km long would break under its own weight.

The range of the observation from the tethered balloon, can be calculated by the formula $l_{obs} \approx (2H_a R_E)^{1/2}$, where H_a – altitude of the balloon, R_E = 6400 km - radius of the Earth. For the altitude of H_a = 7 km the radius of observation is l_{obs} = 300 km, for H_a = 12.5 km the radius of observation is l_{obs} = 400 km. It makes possible to strive for a higher altitude of the tethered balloon with a useful load.

With increasing the altitude of the tethered balloon, besides increasing the mass of the cable holding the balloon the density of the atmosphere sharply exponentially decreases. This decrease of density ρ is usually described by the barometric formula $\rho = \rho_0 \exp[-z/H_0]$, where $\rho_0 = 1.3 * 10^{-3}$ g/cm^3 - air density under normal conditions (on the surface of the Earth), H_0 = 7 km - the height at which the air density decreases by factor e = 2.72 - base of the natural logarithm.

It is not enough to simply raise the balloon to the appropriate height, it is necessary to place a useful load whose mass will be considered to be equal to: m_{load} = 1 ton.

We choose the altitude for the balloon H_a = 7 km and calculate its parameters for this case.

1. Weight of the steel cable

We assume that the density of the cable wire is close to the density of iron, then the mass of the cable wire is equal to 800 kg / km * 7 km = 5.6 tons.

2. Mass of hydrogen

The density of hydrogen is by 14 times less than the density of air. At the height of H_a = 7 km the air density is 4.8 * 10^{-4} g/cm^3, the buoyancy force for a balloon filled with hydrogen is 0.44 kg/m^3. Let the total capacity of the balloon be 10 ton, then its volume should be: V_a = 2.3 * 10^4 m^3.

The density of the hydrogen filling the balloon is:
ρ_{H2} = (1300 g/m^3) / (2.72 * 14) = 34 g/m^3, so the hydrogen mass is equal to:
$m_{H2} = \rho_{H2} * V_a$ = 800 kg.

Assuming that the balloon has a spherical shape , we find that the radius of the sphere must be R_{sp} = 18 m, diameter of the sphere is of the order 36 m, which corresponds approximately to the height of 12 storey building.

3. Obtaining the hydrogen

We consider the production of hydrogen by means of water electrolysis. As it is known one gram – mole of gas under normal conditions, takes up the volume of 22.4 liters.

The mass of one molecule of water is by 9 times greater than the mass of the hydrogen molecule, so, to obtain the amount of hydrogen of $m_{H2} = 800$ kg, it is required to decompose 7.2 tons of water into oxygen and hydrogen.

The energy to decompose one mole of water into oxygen and hydrogen is approximately equal to: $\Delta E = 60$ kJ / mol. The volume of water with a mass $m_{water} = 7.2$ ton contains $4 * 10^5$ moles, and to decompose the water into oxygen and hydrogen it is required to expend the energy $E = 6*10^4 * 4*10^5 = 2.4 \ 10^{10}$ J.

Faraday's first law declares that the mass of the substance released at the electrode is proportional to the current flowing through the electrolyte. The mass is also proportional to the time of the current flow through the electrolyte: $m = k * I * \tau$. Since the hydrogen molecule is diatomic then the flow of the current going via the electrolyte is equal to $I = 2$ A for the time $\tau = 1$s. So, at the exit of the cathode there will be $6 * 10^{18}$ molecules of hydrogen.

One mole contains $6 * 10^{23}$ molecules; the total number of moles in the water mass of 7.2 tons is $4*10^5$, i.e. the volume comprising $2.4 * 10^{29}$ molecules. So, it will be needed for the current to pass through the cell during such a period of time to exceed the charge required for decomposition of one mole of water in $4 * 10^{10}$. This may be achieved, for example, by passing current $I = 2$ MA for time $\tau = 4 * 10^4$ seconds, or during 12 hours.

4. The balloon shell weight

We suppose that the balloon shell is made of a polymer film with a density of $\rho_b = 1.2$ g/cm^3, having a thickness of $\delta_b = 25$ μ. Then 1 m^2 of this film has a mass of: $m_b = 30$ g. The balloon shell square is: $S_b = \pi \ d_b^2 = 3.14 * 36*36 = =4 * 10^3$ m^2. So, the mass of the envelope is equal to: $M_a = m_b * S_b = 120$ kg.

5. The sleeve supplying the balloon with hydrogen

If it is required to fill the balloon with a volume $V_b = 2.3 * 10^4$ m^3 at a pressure by e times lower than the normal pressure, it means that under normal pressure it is required approximately 10^4 cubic meters of the hydrogen to fill the balloon during $4 * 10^4$ seconds, i.e. the desired filling rate for the balloon is of the order of 0.25 m³ / s.

Let the cross- section of the sleeve be of $S_{sleeve} = 0.05$ m², and the rate of gas flow through it is $V_{H2} = 5$ m / s. Then the diameter of the sleeve will be of the order: (π d²$_{sleeve}$ / 4 = 500 cm²), $d_{sleeve} = 25$ cm. The perimeter of the sleeve is about 0.8 m and the square from which the sleeve was made will be equal to $0.8 * 7000 = 5600$ m². We assume that the sleeve supplying the balloon with hydrogen, is made of the same plastic as the balloon, i.e., its surface density is $\rho_{plas} = 30$ g/m². In this case, the mass of the supplying sleeve is equal to $m_{sleeve} = 168$ kg.

6. Transmission of electricity to the balloon

We calculate the resistance of the copper wire with a cross section $s_{line} = 1.7$ mm² and a length of 7 km from the formula:

$$R = \rho \ l/s_{line} = 1.7 * 10^{-6} * 7 * 10^5/(1.7 * 10^{-2}) = 70 \ \Omega. \qquad (1)$$

It can be seen that the direct current electric resistance is sufficiently high and the transfer of energy to the useful load can cause a problem.

The mass of the copper wire is: $m_{Cu} = 9 * 1.7 * 10^{-2} * 7 * 10^5 = 100$ kg.

Let the load resistance be equal to $R_{load} = 10 \ \Omega$ and the applied voltage is $U = 400$ V. Then the current flowing in a circuit will be equal to $I = 5$ A and From the total power of 2 kW this current will allocate power $P = 1750$ W in the supplying wires and only 250 W - in the useful load.

If you increase the cross section of the wires supplying with electricity by the order of up to 17 mm², then the supplying wire electricity resistance will drop to 7 Ω. If the useful load resistance is equal to 73 Ω, then at the same current in the circuit $I = 5$ A, and the total power of 2 kW, the highlight power $P_{load} = 1825$ W remains in the useful load and only $P_{cab} = 175$ W will go to the supplying wires.

The mass of the wires supplying the electricity will increase to $m_{Cu} = 1$ ton.

7. Wind loading

Now we solve a task how to hold the balloon at a predetermined altitude. When the diameter of the balloon $d_b = 36$ m, its capacity is 10 ton. From these ten tons, 5.6 tons are the mass of the steel cable, we have taken the mass of the useful load to be equal to 1 ton, 0.12 ton is the mass of the balloon shell, 0.17 tons is the mass of the sleeves supplying hydrogen to the balloon. The mass of the supplying of wires is equal to: $m_{Cu} = 1$ ton. After summing up all the loads we obtain the resulting force of approximately 7.6 tons, which will be directed upwards.

The durability of the steel cable against breaking is equal to $\sigma_{st} = 340$ kgf/mm^2 or 34 tons/cm^2 [2], p. 55, so that the permissible additional force acting on the cable upward is equal to 20 tons.

The pressure composed by the wind is: $P_w = C_x \rho V_w^2 / 2$. Considering that for our range of the wind velocity $V_w \approx 10$ m / s, the drag coefficient C_x of the balloon, is: $C_x \approx 0.5$, we find that for wind velocity $V_w = 10$ m / s the pressure of the wind will be:

$$P_w = 0.5 * 5 * 10^{-4} * 10^6 / 2 = 1.25 * 10^2 \text{ dn/cm}^2. \tag{2}$$

For the transverse cross-section of the balloon $S_{cb} = \pi d_b^2 / 4 = 10^7$ cm^2, the force will be equal to: $F_w = 1.25 * 10^9$ dn or 1.25 tons.

It is evident that the allowed gusts of the wind going upward can be of velocity: $V_w = 40$ m / s.

8. Lateral displacement of the balloon caused by the action of the side wind

If the balloon is tied with one cable, its deviation caused by the action of side winds can be calculated from the following considerations.

Considering the deviation length of the cable to be equal to 1, we obtain that the length of the non-deviated cable is equal to cos α. The relative extension for small angles α is equal to $\Delta H_a / H_a = (1 - \cos \alpha) \approx \alpha^2 / 2$. The reverse force for the balloon is equal to the tension force multiplied by sin α, so that the equation relating the deviation angle with the wind force looks as follows:

$$F_w = F_{back} \approx E * S * \alpha^3 / 2, \tag{3}$$

63

where: $E = 2 * 10^{12}$ dn/cm^2 - Young's modulus for the iron, $S = 1$ cm^2 – is the transverse cross-section of the cable.

For the wind force $F_w = 1$ ton, at the wind velocity 9 m / s we find that angle α is equal to α = 0.1 and the linear deviation of the balloon will be as follows: $\Delta H_a = H_a * \alpha = 700$ m.

We fix additional three cables with a cross-section $S_{cab} = 0.1$ cm^2 each along the perimeter of the balloon with the inclination angle to the horizon φ = 45^0. The length of each of the cables will be equal to $l_{cab} = 10$ km, the extension of the cable under the force of the wind directed along the cable can be found from Hooke's law:

$$\Delta l_{cab} / l_{cab} = F_w / E * S_{cab}. \qquad (4)$$

From equation (4) it follows that for the wind force $F_w = 200$ kgf (the wind velocity 4 m / s) the extension of the cable is equal to: $\Delta l_{cab} = 10^{-3} l_{cab} = 10$ m.

Conclusion

It is evident that the durability holding the balloon is enough to withstand the winds of considerable strength. The main problem for such a tethered balloon is its displacement under the action of the side wind loading.

Literature

1. Tables of Physical Quantities, Handbook ed. I. K. Kikoin, Moscow, Atomizdat, 1976

2. Tables of Physical Quantities, Handbook ed. I. S. Grigor'ev and E. Z. Meylikhov, Moscow, Energoatomizdat, 1991

Gas-dynamic acceleration of bodies till the hyper sonic velocity

The article considers an opportunity of gas-dynamic acceleration of body from the initial zero velocity till the finite velocity: V_{fin} = 5 km / s. When the gas flow rate of the body pre-acceleration reaches V_{in} = 1 km / s, the body is accelerated at the front of the explosion wave propagating along the coils of the hexogen spiral. This wave accelerates the body and, finally, it reaches the velocity of 5km/s. The accelerated body has mass m = 0.1 kg and diameter d_{sh} = 11.3 mm. Acceleration length is L_{acc} = 6 m. At the slope of the spiral to the horizon equal to $\Theta = 70^0$ the flight range of the body is equal to: S_{max} = 1600 km, and the maximum height of the flight is H_{max} = 1100 km.

Introduction

There is a known [1] method of gas-dynamic acceleration of the body in the trunk. The powder disposed in the trunk for a short period of time transfers from the solid to the gaseous state, after that; due to its expansion it pushes the body out from the trunk. The finite velocity of the body in this method of the gas-dynamic acceleration is determined by the density and temperature of the composed gas.

Increasing the density of the gas and its temperature is limited by the durability of the trunk and can not be increased. Modern systems are designed for pressure 3500 atm. and temperature 3000 C^0. Further increase of the temperature and pressure beyond these values is not justified because at high temperatures the trunk is damaged and increase of the pressure requires the quality of the trunk to be improved. Further, at high temperatures there is an opportunity of dissociation of molecules of the gas that is related with the loss of the thermal energy of the gas. Therefore, at the present state of the technique development the achieved velocity of the body is the upper limit.

Taking into account the heat dissipation losses due to friction against the walls of the gas trunk and other losses it is possible to accept the limited velocity approximately equal to $V \approx 2$ km / s. This velocity is achieved when the body mass is a negligibly small part of the mass of the gunpowder. If you increase the mass of the body, the body velocity will seek to reach the ordinary velocity of the body: $V_{ord} \approx 1$ km / s.

To obtain the hypersonic velocity of the body about $V_{sh} \approx 5$ km / s is impossible by using this method.

In principle, one can accelerate the body by the explosion wave starting with the zero initial velocity of the body. At certain synchronization of the body with the explosion wave it is possible to accelerate the body at non- zero initial velocity. However, this method has a fundamental drawback: it is the small length interaction of the body with the explosion wave.

Let us consider an opportunity of accelerating the body till the hyper sonic velocity formed in the space by the explosion wave propagating synchronically with the accelerated body.

Selection of basic parameters

As the explosive substance we have chosen [2] the hexogen cord with a diameter $d_{cord} = 0.5$ cm, which is wound onto the tube with a diameter $2r_0 = 3$ cm. Let the body which is supposed to accelerate, be of a cylindrical shape with mass $m_b = 0.1$ kg, cross-section $S_{tr} = 1$ cm^2, and it has the initial velocity $V_{in} = 1$ km / s. The finite velocity of the body is chosen to be equal to $V_{fin} = 5$ km / s and the acceleration length is equal to $L_{acc} = 6$ m.

To obtain this acceleration, we find the following from the equation of motion of the body:

$$m_b dV / dt = F_{axis}, \tag{1}$$

that the accelerating force must be equal $F_{axis} = 2 * 10^5$ N $= 2 * 10^5$ kg $*$ m/s^2. Indeed, dividing this force by body mass $m_b = 0.1$ kg, we find that the body will move with the uniformed acceleration: $a = 2 * 10^6$ m/s^2.

The velocity of the body will increase in accordance with the law:

$$V = V_{in} + at, \tag{2}$$

and from the initial velocity $V_{in} = 1$ km / s till the finite velocity $V_{fin} = 5$ km / s the body will be accelerated during $t_{acc} = 2$ ms.

To implement the impact the permanently acting force onto the body, it is necessary to make an explosion of the hexogen cord wound on the carcass as a double spiral structure. The winding step must permanently grow so that the wave propagation velocity along the spiral would permanently coincide with the velocity of the body.

The dependence of the passed distance (at the uniformed accelerated motion) on the time can be written as follows:

$$L = V_{in}t + at^2 / 2 . \tag{3}$$

Solving this equation for the velocity, we find that the dependence of the velocity on the covered distance is expressed as follows :

$$V(L) = (V_{in}^2 + 2\ aL)^{1/2} . \tag{4}$$

At the initial parameters, when $L = 0$, the velocity of the body is equal to $V = V_{in} = 1$ km / s. At the finite parameters: $L = L_{acc} = 6$ m, the velocity of the body is equal to: $V = V_{fin} = 5$ km / s.

Knowing the law of the velocity growth depending on the length of the passed distance, it is possible to find the dependence of the winding step on the passed distance.

Assume the velocity of propagation of detonation in hexogen to be equal to $V_{det} \approx 8$ km / s [2]. To have the velocity of the explosion wave V_{axis}, propagating along the axis of the spiral, permanently coincide with the velocity of the body V_b, the following relation must be satisfied:

$$V_b = V_{axis} \approx V_{det}h/2\pi r_0, \tag{5}$$

where V_b - velocity of the body, V_{axis} - the velocity of the explosion wave propagating along the axis of the spiral, r_0 - radius of the spiral winding, h - the winding step of the spiral.

Detonation propagating along hexogen cord runs around the perimeter of the spiral, along the axis it runs with velocity $V_{axis} \approx V_{det}h/2\pi r_0$. Hence, for the initial velocity of the body, $V_{in} = 1$ km / s, the winding step should be equal to: $h_{in} \approx 2\pi r_0 V_{in}/V_{det} = 1.17$ cm, the finite step of winding of the spiral must be equal to: $h_{fin} \approx 2\pi r_0 V_{fin}/V_{det} = 5.88$ cm. Intermediate values of the winding step of the spiral are given by the following ratio:

$$h(L) = V(L) * V_{det}/2\pi r_0. \tag{6}$$

We will consider the detonation of the double spiral with a diameter of each of the hexogen cords equal to d_{cord}, equal to: $d_{cord} = 0.5$ cm.

Pressure on the detonation front in hexogen reaches $P_{hex} \approx 30$ GPa [2]. Let the body be located at the distance equal to the radius of the spiral winding $z = z_s = 1.5$ cm at the moment when the explosion wave comes up to its back slice.

We will consider the detonation of the long cylindrical hexogen cord. In the spread out of the explosion products the pressure of the explosion wave P_{wave} at the back slice of the body would have been less than the pressure on the front of the wave by the ratio of the square of the radius of the hexogen cord: $(d_{cord} / 2)^2 = 0.625$ cm^2 to the square of the distance till the back slice of the body (4.5 cm^2). It would have been as follows: $P_{wave1} = P_{hex} r^2_{cord} / (r_0^2 + z_s^2) = 0.4$ GPa. Since the construction of detonating chamber is like this, the explosion products leave the area of the explosion only through one quadrant, the pressure at the front of the explosion wave is by 4 times more than in the free space and equal to $P_{wave2} = 1.6$ GPa.

In this case the force acting on the back slice of the body is as follows:

$$F_{axis} = P_{vawe2} * S_{tr} * \cos \varphi, \qquad (7)$$

where we have chosen : $\varphi = 45^0$, $\cos \varphi = 0.7$. Taking into account that two cords are simultaneously detonated, their forces are summed up.

Substituting the numerical values into formula (7) we find that this force is equal to $F_{axis} \approx 2.2 * 10^5$ N.

The light carcass with the hexogen spiral and rings of the special shape, as well as the separable cylinder are located inside the strong trunk with the internal diameter of $D_{bar} = 80$ mm. In this case the pressure onto the inner surface of the trunk at the absence of the substance between the strong trunk and hexogen cord would have been equal to the value: $P_{bar} = P_0 * (d_{cord} / 2)^2 / [(D_{bar} / 2) - r_0]^2 \approx 3000$ atm., which is the ordinary pressure for the guns [1]. Due to the presence of the rings and separable cylinder the pressure on the inner surface of the strong trunk will be less.

After reaching the velocity $V = 2$ km / s it is possible to use a quadrifilar spiral. But it is necessary to reduce the diameter of the two cords accordingly (by the root of 2 times). When velocity $V = 3$ km / s it will be possible to use the six-filar spiral, and etc.

Synchronization of the body being accelerated and the explosion wave

As it is shown in [3], the velocity of detonation product spread out of the cylindrical cord in the transverse direction is: $V^{\perp} = 0.8 V_{det}$, i.e. In our case this velocity is approximately equal to $V^{\perp} \approx 6.5$ km / s. Detonation products should reach the back slice of the body at the moment when the back slice is at the distance equal to the radius of the spiral - $l_{in} = r_0 = 1.5$ cm. Then the distance from the back slice of the body to the explosive area of the detonating cord must be $l_{body} \approx 2.12$ cm.

Explosion products fly over this distance during the time τ_{shok} equal to: $\tau_{shok} = l_{body} / V^{\perp} = 3.26$ μs. The body moving with the initial velocity $V_{in} = 1$ km / s, must be at this moment at the distance $l_{in} = 3.26$ mm from the start of the acceleration.

Phase stability principle for the body being accelerated on the front of the explosion wave

As in particle accelerators on the traveling wave there is the phase stability [4] while accelerating the body at the front of the explosion wave. In the particle accelerators the acceleration phase for particles is chosen in advance, it is called synchronous. For this phase the longitudinal motion of the particles is calculated after that. If any particle is accidentally behind the synchronous phase, it will get into the stronger electric field. Then it will be more accelerated and finally will catch up with the synchronous phase.

If the particle is too fast in its motion for the synchronous phase, then it will get into a smaller electric field, it will be less accelerated, and, finally, the accelerated moving pulse and the synchronous phase will catch up with the particle.

Below we show that in this case while accelerating the body at the front of the explosion wave, the dependence of the force acting for the body on the phase of the pulse at its front, has a declining character.

The force acting on the body, $F_{axis} = P_{wave} * S_{tr} * \cos \varphi$, depends on the pressure $P_{wave} = P_{hex} * r_{cord}^2 / (r_0^2 + z^2)$, where $P_{hex} = 30$ GPa - pressure of the explosion wave (hexogen), $2r_{cord} = 0.5$ cm - diameter of the hexogen cord, $r_0 = 1.5$ cm – the hexogen cord winding radius, z - the distance along the axis from detonating area till the back slice of the body (cm). Projection of the pressure force on the axis of the acceleration is proportional to cos φ, which can be represented as follows: $\cos \varphi = z / (r_0^2 + z^2)^{1/2}$.

Thus, the dependence of the force acting on the body along the axis (F_{axis}) on the distance along the axis between the detonating area and the back slice of the body, is expressed as follows:

$$F_{axis} \sim z / (r_0^2 + z^2)^{3/2}. \tag{8}$$

Below we compile a table of the values of this function for three values of the distance between the detonation region and the back slice of the body.

Table 1. Values of the function $z / (r_0^2 + z^2)^{3/2}$

z, cm	$z/(r_0^2+z^2)^{3/2}$
1	0.17
1.5	0.157
2	0.128

From the analysis of this function it is clear that the force acting on the body reduces if the body is faster than the synchronous phase. The explosion force acting on the body increases if the body is behind the synchronous phase. This dependence corresponds to the stable phase of the longitudinal motion according to the phase stability principle.

Transverse movement of the body

Let us consider the transverse force available at the front of the explosion wave on the body. In the beginning of the acceleration the velocity of the explosion wave propagating along the axis of the spiral is $V_{in} = 1$ km / s. The detonation velocity propagating along the spiral is $V_{det} = 8$ km / s. The perimeter of one of the spiral coils is $\pi d_{spir} \approx 10$ cm, so that the time of detonation of the total cycle is as follows: $\tau_{cyc} \approx \pi d_{spir} / V_{det} = 10$ μs.

Assume that the body moves transversely accelerated but by two orders of the magnitude lower than in the longitudinal direction, i.e., $a = 2 * 10^4$ m/s². During the time when the detonation runs one third of the total cycle $\tau_{cyc1/3} = 3$ μs, the body is shifted by a distance of $S_{1/3} = a \tau_{cyc\ 1/3}^2 / 2 = 0.1$ μ. After that the transverse force changes its direction and after the total cycle its action is averaged.

Ballistics. Aerodynamic resistance

Now we calculate the motion of the body released at an angle of $\Theta = 70^0$ to the horizon, taking into account the air resistance. The equation of the horizontal motion of the body can be written as follows:

$$m_b \, dV_x / dt = \rho C_x S_{tr} V_x^2 / 2, \tag{9}$$

where m - mass of the body , V_x- horizon velocity, g - 0.01 km/s^2 - acceleration due to gravity , $\rho = \rho_0 e^{-z/H0}$ - barometric formula of the change of the atmospheric density in dependence on the height , $\rho_0 = 1.3 * 10^{-3}$ g/cm^3 - air density at the Earth surface, $H_0 = 7$ km - the height at which this density decreases by a factor of e.

The aerodynamic coefficient or coefficient of drag resistance is presented as a dimensionless quantity, taking into account the "quality" of the body shape:

$$C_x = F_x / (\tfrac{1}{2}) \, \rho_0 V_x^2 S_{tr}. \tag{10}$$

The solution of equation (9) can be written as follows:

$$V(t) = V_x / [\, 1 + \rho C_x V_x * S_{tr} * t/2m_b \,]. \tag{11}$$

In order to calculate the change in velocity of the body in dependence on time, you need to find the aerodynamic coefficient C_x.

Calculation of the aerodynamic coefficient for the air

We assume that the body has the shape of a cylindrical rod with a conical head. Then at the hit of a nitrogen molecule on the sharp cone, the change of the longitudinal velocity of the molecules is equal to:

$$\Delta V_x = V_x * \Theta_t^2 / 2, \tag{12}$$

where Θ_t –the cone angle at the vertex. Gas molecules transfer the momentum to the body:

$$p = mV = \rho V_x S_{tr} t * \Delta V_x. \tag{13}$$

The change in the momentum per unit of the time is the force which is called the force of the frontal slow down,

$$F_{x1} = (\tfrac{1}{2}) \rho V_x S_{tr} * V_x * \Theta_t^2. \qquad (14)$$

Dividing F_{x1} by $(\tfrac{1}{2}) \rho V_x^2 S_{tr}$, we obtain the drag coefficient for the sharp cone at the mirror reflection of the molecules from the cone, (Newton's formula):

$$C_{x\,air} = \Theta_t^2. \qquad (15)$$

Our consideration corresponds to hypersonic velocity; we can neglect the effects that occur at the velocity close to the sonic velocity in the unperturbed medium. Let the length of the conical part of the body be as follows: $l_{cone} = 65$ mm and the diameter of the body $d_b = 11.3$ mm. This means that the angle at the vertex of the cone is: $\Theta_t = 0.173$ and $C_{x\,air} = 0.03$.

Substituting this value in the formula for the C_x loss of the longitudinal velocity in dependence on time (11), we find that the decrease in the vertical velocity of the body in the first second of the flight is of the order of 10%.

The flight range of the body is $S_{max} = 2V_0^2 * \sin\Theta * \cos\Theta / g = 1600$ km, the maximum lifting height of the body is $Y = V_0^2 * \sin^2\Theta/2g = 1100$ km.

Fig. 1 shows a diagram of the device.

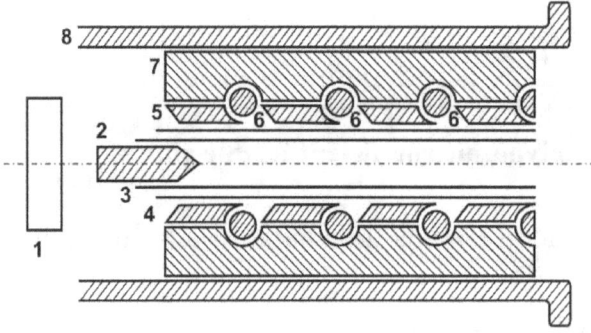

Fig.1. 1 – the gun, 2 - the body, 3 – the light trunk, 4 - light carcass, 5 - special shaped rings, 6 - hexogen cord, 7 - separable cylinder, 8 - solid trunk.

The operation of the device is as follows. In the gun (1), the acceleration of the body (2) of a cylindrical shape with mass $m_b = 0.1$ kg, and the cross-section of $S_{tr} = 1$ cm^2 is performed by using the ordinary gas - dynamic method. The body is centered relatively the spiral with a light trunk (3). The body moves in the light carcass (4), where the special shaped rings (5) are embedded. Synchronously with the start of the acceleration we produce the explosion of the hexogen cord (6), which is placed in the gaps between the rings. Outside the cord is surrounded with the separable cylinder (7). The rings must be rigidly fixed with the separable cylinder. Figure 1 does not show fastening of the rings to the cylinder. All the assembly is placed inside the solid trunk (8). Detonation propagates along the cord and creates the explosion wave running synchronously with the body.

If to increase the mass of the body by an order of the magnitude till the value of $m_{b1} = 1$ kg, then the length of the acceleration (at the same parameters) will increase by the order of magnitude and will be equal to the value of $L_{accl} = 60$ m. Such an accelerator may be arranged only horizontally, while at the same time the body must have asymmetry which creates the lifting force.

Suppose that the lifting coefficient is, $C_y = 0.2$. In this case the drag coefficient must be the same: $C_x = 0.2$.

Thus, the equation of the vertical motion can be written as follows:

$$m_{b1}dV_y / dt = C_y \, \rho_0 V_x^2 * S_{tr} / 2, \tag{16}$$

where C_y is the aerodynamic lifting force coefficient, $\rho_0 = 1.3 * 10^{-3}$ g/cm^3 - the air density on the surface of the Earth, $V_x = 5$ km / s - the horizontal velocity of the body, S_{tr} is the cross-section of the body.

Solving approximately equation (16), we obtain:

$$V_y = C_y \rho V_x^2 S_{tr} t / 2 m_{b1}. \tag{17}$$

Integrating again, we obtain an expression for the lifting height of the body:

$$Y_1 = C_y \rho \, V_x^2 \, S_{tr} t^2 / 4 m_{b1}. \tag{18}$$

Solving the equations of horizontal and vertical movements we can obtain the time dependence of the flight parameters of the body, which we present in Table 2. The first column gives the time of the flight, the second - the horizontal velocity, in the third – there is the vertical velocity, the fourth column shows the height of the lifting of the body.

Table 2. Flight parameters for the case of C_x, $C_y = 0.2$.

t, s	V_x, km/s	V_y, km/s	Y, km
0	5	0	0
2	4.43	0.65	0.65
5	3.8	1.37	2.7
10	3	2.3	8.5
20	2.68	2.66	20

The time of the body lifting till the maximum height in this case is as follows: $\tau_{max} = V_y / g = 266$ s. The flight range of the body is $S_{max1} = V_x * 2\tau_{max} = 1400$ km, the maximum body lifting height is $Y_1 = V^2{}_y/2g = 350$ km. Changing the shape of the cone in the head part of the body, it will be possible to obtain different trajectories.

Conclusion

These flight parameters may be of interest for a number of applications, in particular, for the removal of the space garbage at the low orbits.

Literature

1. I. A. Sterznev, Artillery guns multiple actions, the speed limit of artillery shells, http://ivanstrezhnev.appspot.com/3/3.html

2. http://ru.wikipedia.org/wiki/Гексоген

3. V.V. I'lyin, A.P. Rybakov, V.V. Kozlov, Mathematical model of expansion of explosion products at outlet oblique detonation wave at the free surface, Electronic scientific journal "INVESTIGATED IN RUSSIA", 2006, p. 1531, http://zhurnal.ape.relarn.ru/articles/2006/165.pdf

4. Veksler, Reports of the USSR, v. 43, issue 8, p. 346, 1944
   E. M. McMillan, Phys. Rev., V. 68, p. 143, 1945

Artificial ozone holes

This article considers an opportunity of disinfecting a part of the Earth surface, occupying a large area of $\sim 10^4$ km^2. Such a need may arise, for example, in the case of uncontrolled spreading out of different viruses or other infections. The method of disinfection being described in this article is more close to the nature than the radiation sterilization.

Decontamination of material and objects by means of this method will be carried out by using strong ultraviolet solar radiation through an artificial ozone hole in the Earth atmosphere. To realize this method, it is necessary to launch a balloon into the atmosphere, higher than the upper limit of the distribution of ozone into the atmosphere to the height of $H_1 = 30$ km. In the balloon gondola there is bromine in the solid state. The sunlight will cause dissociation of molecular bromine into atoms, each bromine atom kills $M_{Br + O3} = 3 * 10^4$ molecules of ozone. Each bromine plate has a mass of $4 * 10^{-2}$ grams and destroys ozone in the area of 10 x 10 meters. Thus, to form the ozone hole over the area of 10 thousand square kilometers, it is required to have the total mass of bromine equal to the following: $m_{Br\ 1} = 4$ tons.

Introduction

The main source of ultraviolet radiation on the Earth is the Sun. The Sun radiation power is much stronger than all the existing light sources on the Earth. Although the power density of solar radiation in the hard ultraviolet is rather small [1], due to the fact that materials and objects can be irradiated on the area of thousands of square kilometers, the total power can be hundreds of Giga Watts, that is impossible to have with mercury lamps.

However, almost all hard ultraviolet radiation of the Sun is absorbed by the ozone layer of the Earth [2]. The natural ozone holes (thinning of the ozone layer), are dangerous because in the absence of ozone the hard ultraviolet radiation of the Sun penetrates through the Earth atmosphere and produces undesirable sterilization of materials and objects.

Ozone is effectively decomposed by bromine in the following chemical reactions:

$$Br + O_3 \rightarrow BrO + O_2 \quad \text{и} \quad BrO + O \rightarrow Br + O_2.$$

In the result of these reactions, the molecules of ozone O_3 are converted into oxygen molecules O_2, but initial atoms of bromine remain in the free state and again participate in this process. Every atom of bromine destroys ozone molecules: according to [3], this number is equal to a million (10^6), the other data [4] give the number equal to 10^5 ozone molecules. After that it escapes from the atmosphere in the result of other chemical reactions. Let us optimistically assume that the multiplicative parameter M_{Br+O3}, in this reaction is equal to: $M_{Br+O3} = 3 * 10^4$.

We remember that atomic bromine is obtained from dissociation of Br_2 under the influence of sunlight.

1. The drift of bromine molecules in the air under the action of gravity

We consider the process of bromine molecules coming down to the Earth under the action of gravity after their spraying out in the stratosphere above the region occupied by the ozone layer. Besides the gravity the molecules of bromine are influenced by the resistance related with their friction with the air.

We find the drift velocity from the following equation:

$$m_{Br}\, dV_z/dt = m_{Br}g - C_z S_{tr\,m}\, \rho V_z^2/2, \qquad (1)$$

where m_{Br} is the mass of the molecule of bromine, V_z - vertical velocity, $g = 10^3$ cm / s^2 is the acceleration of gravity, C_z - drag coefficient, $S_{tr\,m}$ – transverse cross-section of the molecule of bromine.
The $\rho = \rho_0 * \exp[-z / H_0]$ – is a barometric formula, where $\rho_0 = 1.3 * 10^{-3}$ g / cm^3 which is the air density under normal conditions, $H_0 = 7$ km, it is the altitude at which the air density decreases by factor e.

Assuming the established velocity of bromine atoms to be equal to $dV_z / dt = 0$, we find from (1)

$$V_{z\,drift} = (2m_{Br}\, g/C_z S_{tr\,m}\, \rho)^{1/2}. \qquad (2)$$

After substituting the numerical values: $m_{Br} = 160 * 1.6 * 10^{-24}$ g, $C_z = 1$, $S_{tr} = 10^{-16}$ cm^2 that is the transverse cross-section of the molecule of bromine, $\rho = 2.6 * 10^{-5}$ g / cm^3, we find that the drift velocity, i.e. – the velocity at which the molecules of bromine are falling to the Earth from the height $H_1 = 30$ km, is: $V_{z\,drift} \approx 10$ cm / s.

2. Diffusion of bromine molecules in the radial direction

Consider the process of "spreading" the gaseous bromine cloud over the radius while it's moving slowly downwards under the influence of gravity. The cloud will increase its size due to diffusion.

The diffusion process will go on the radius according to Einstein's formula:

$$r^2 = D * t, \tag{3}$$

where D is the diffusion coefficient for gases equal to: $D = (1/3) l_a * \bar{v}$, l_a - the mean length of free passing, v is the average velocity of bromine molecules, r – the radial coordinate, t - time.

Define, first, the mean free passing by using the following formula:

$$l_a = 1/n\sigma, \tag{4}$$

where n is the number of molecules of air in cubic centimeters at a given height, σ – their collision cross section.

According to the barometric formula, the number of molecules in a cubic centimeter at a height $H_1 = 30$ km, will be by 50 times less than the number Loschmidt, i.e., the number of molecules in a cubic centimeter under normal conditions. Taking into account the Loschmidt number equal to: $n_0 = 2.7 * 10^{19}$ molecules / cm^3, we find $n = 5.4 * 10^{17}$ molecules / cm^3. As the interaction cross section we can take the transverse cross-sectional of the molecule of bromine, which was taken to be equal to: $S_{tr\,m} = 10^{-16}$ cm^2. Then, from formula (4), $l_a = 2 * 10^{-2}$ cm.

Now we define the quantity $\bar{v}$ - the average velocity of bromine molecules. We find it from the following relation:

$$m\bar{v}^2/2 = (3/2)k * T, \tag{5}$$

where $k = 1.38 * 10^{-16}$ egr / degree - Boltzmann constant, $T = 300$ ^{0}K- gas temperature. Then, according to the formula (5), the average velocity of the molecules of bromine is: $\bar{v} = 3 * 10^4$ cm / s.

You can now find $D = (1/3) l_a * \bar{v}$, - diffusion coefficient of bromine molecules, which in our case is equal to: $D = 2 * 10^2$ cm^2 / s.

The average radius of the spreading out the bromine molecules in space will grow with time as:

$$r = (D*t)^{1/2}. \tag{6}$$

It can be seen that the characteristic time: $t_{10} = 10^5$ s, during which the bromine molecules will drift down from the height of $H_1 = 30$ km at a distance $\Delta H_1 = 10$ km, along the radius the bromine cloud will spread out at a distance of: $r_{10} = (D * t)^{1/2} = 40$ meters.

This means that if we want to uniformly fill a large area (100 * 100 km), then the initial bromine clouds must be placed from each other at a distance of not more than $\Delta r_{in} < 10$ meters.

3. Uniform distribution of crystallites of bromine over a large area

We consider how to uniformly fill a large area with gaseous bromine. We take the case when the bromine presented in the form of small solid crystals is thrown out by means of a centrifuge, similar to the way as the road machines throw sand onto the roadway.

The bromine crystals should fly quite a long distance in the radial direction - about 50 kilometers before reaching the desired point. Having reached it they have to melt, warm up till the evaporation temperature and evaporate. The distance between the neighboring crystals should be not more than 10 meters for after their evaporation all the required area would be uniformly filled with gaseous bromine.

3.1. Quantitative relationships

In the column with an area of 1 cm^2 at normal conditions there are 10^{19} ozone molecules, in the column with an area of 1 m^2 their number is 10^{23}. We consider the processes in the standard "box" with dimensions of 10*10 m, in the column with this window there are 10^{25} molecules of ozone.

We assume that one bromine atom decomposes $3 * 10^4$ molecules of ozone. It means that in order to decompose all the molecules in this window it will be required to have $3 * 10^{20}$ atoms of bromine.

At 100% dissociation in the column there must be $1.5 * 10^{20}$ molecules of bromine, taking into account that the bromine molecule consists of two atoms Br.

Let us find the mass of molecular bromine from the relation:

$$6*10^{23} \text{ ------------ } 160 \text{ g}$$
$$1.5*10^{20} \text{ ------------ } x \text{ g,}$$

from where $x = 4 * 10^{-2}$ g, i.e. the mass of a crystal of bromine is equal to $m_{Br} = 4 * 10^{-2}$ g. Densities of solid and liquid bromine differ by less than 1%, [6], p. 188. The density of liquid bromine is: $\rho_{Br} = 3.12$ g / cm³, then the volume occupied by a drop of bromine is equal to: $V_{Br} = 10^{-2}$ cm³.

It is possible to find the ball diameter, assuming that the volume of the ball: $V_{Br} = (4/3) \pi r_{Br}^3 = (\pi / 6) d_{Br}^3 \approx d_b^3 / 2$. From this we find: $d_{Br} = 0.27$ cm. The ball of this diameter contains $1.5 * 10^{20}$ molecules of bromine, which, after dissociation will give $3 * 10^{20}$ bromine atoms and these atoms will be able to kill 10^{25} ozone molecules in a column with a base of 10*10 m.

3.2. Aerodynamics of bromine crystals having a spherical shape

We find the Reynolds number for a ball of solid bromine (the melting temperature of bromine is $T_m = -7.3$ ⁰C) according to the following formula:

$$Re = \rho V_0 l/\eta, \tag{7}$$

where ρ is the density of air at a given altitude, 0 - the velocity of the ball in the air, l - the characteristic length, in our case: $l = d_{Br} = 0.27$ cm, $\eta = 1.8 * 10^{-4}$ Poise - the viscosity of air, [5], p. 273. Substituting numbers for the height $H_1 = 30$ km, and for the initial velocity of the ball $V_0 = 200$ m / s, we find that the Reynolds number: $Re = 7 * 10^2$. This means that the dependence of the resistance on the velocity is linear and the velocity in dependence on the distance will decrease exponentially:

$$V = V_0 \exp[-(\rho C_x S_{tr}/m_{Br})x], \tag{8}$$

where C_x – the drag coefficient of the ball, $C_x = 0.6$, S_{tr} – the transverse cross-section of the ball $S_{tr} = \pi d_{Br}^2 / 4 = 5.7 * 10^{-2}$ cm².

The value shown in the parentheses of the exponent, the so-called "ballistic coefficient" describes the "quality" of the object. The value inverse to the ballistic coefficient has the dimension of length and can be considered as the characteristic length on which the body stops since its velocity decreases by

79

factor e. In our case, the length is equal to:

$x_0 = m_{Br} / \rho C_x S_{tr} = 4 * 10^{-2} / (2.6 * 10^{-5} * 0.6 * 5.7 * 10^{-2}) = 4.6 * 10^4$ cm, or $x_0 \approx 500$ m.

3.3. Possible sizes of the ozone hole and the required mass of bromine for its forming

Suppose you want to form the ozone hole with the dimensions of 100*100 km. This area contains 108 "windows" with the size of 10 * 10 m, in each of them it is necessary to place a crystal of bromine with a mass of $m_{Br} = 4 * 10^{-2}$ g.

Thus, to form the ozone hole with dimensions of 100*100 km, it is necessary to uniformly distribute $m_{Br\ 1} = 4$ tons bromine over this area.

3.4. Opportunities of the balloon

We consider what load capacity to a height of $H_1 = 30$ km has a balloon. Let the diameter of the ball be $d_b = 100$ m. Then, the total volume of such a balloon will be as follows: $V_b = (4/3) \pi r_b^3 = (\pi / 6) d_b^3 \approx d_b^3 / 2 = 5 * 10^5$ m^3.

The air density at the height of $H_1 = 30$ km, is $\rho = 2.6 * 10^{-5}$ t/ m3, so that the load capacity of the empty balloon is equal to13 tons. If the balloon is filled with hydrogen gas, whose density is by14 times less than the density of the air, the balloon load capacity will be by 7% lower than the empty balloon and equal to 12 tons.

Suppose that the balloon shell is made of a polymer film having the density of $\rho_b = 1.2$ g / cm^3, and thickness of $\delta_b = 25$ μ. Then 1 m^2 of such a film has a mass: $m_b = 30$ g. The area of the shell of the balloon is equal to: $S_b = \pi d_b^2 = 3.14 * 10^4$ m^2. The mass of the shell is equal to $m_b * S_b = 1$ ton.

Thus, the load capacity of such a balloon is equal to 11 tons; 4 tons of them are the bromine mass. The remaining mass may consist of auxiliary equipment: a cooling chamber, the centrifuge needed for spreading out the crystals of bromine, energy supply, navigation, control, and so on.

3.5. Aerodynamics of the bromine crystals having the form of thin plates

It follows from (8), that the range of the flight of the bromine crystals in the Earth atmosphere at a height of $H_1 = 30$ km, is not great and is about 500 m.

For uniform distribution of the bromine crystals on the surface with dimensions of 100 * 100 km, the flight distance must be greater than 50 km.

This distance of throwing out the bromine crystals can be obtained if they are formed as cylinders having a sharp cone at the head end.

Indeed, the velocity of bromine crystals chosen by us: $V_0 = 200$ m / s, is $M = 0.6$ Mach units, where $M = 1$ corresponds to the velocity equal to the velocity of the sound in air under normal conditions: $V_s = 330$ m / s. For a ball, for this velocity, the drag coefficient is $C_x = 0.6$, [6].

For a cylinder with a sharp cone in the head, according to the empirical formula [7], the drag coefficient is equal to:

$$C_x = (1.56 + 1.95/M^2) \Theta_{1/2}^{1.7}, \qquad (9)$$

where $M = 0.6$ - Mach number, Θ - the half-angle of the cone at the vertex. As the full angle at the vertex of the cone, it is possible to obtain the relation: $C_x = \Theta^2$, Newton's formula. It is seen that at a small angle of the cone, the drag coefficient may be very small.

To have the distance of throwing the crystals of bromine to be about 50 kilometers, according to the formula (8) the drag coefficient should be equal to: $C_x = 5 * 10^{-3}$, or the angle at the vertex of the cone $\Theta_s = 7 * 10^{-2}$.

If we imagine a bromine crystal as a cylinder with a mass $m_{Br} = 4 * 10^{-2}$ g, having a volume: $V_{Br} = 10^{-2}$ cm^3, its total length is equal to: $l_t = 3$ cm, diameter $d_{Br\,1} = 0.7$ mm and length of the cone part $l_c - 1$ cm. The angle at the vertex of the cone $\Theta_s = d_{Br\,1} / l_c$ if to substitute the numbers it will be equal to what we expected: $\Theta_s = 7 * 10^{-2}$.

3.6. Phase transitions: solid body-liquid and liquid - vapor

The bromine crystal must not only reach the given point, but it must reach it in a gaseous state. For bromine the phase transition energy from solid to liquid is $\Delta Q_s = 10$ kJ / mole, [8], 289.

Assuming 1 mole to be equal to160 g, we find that the energy required to melt the crystal of mass $m_{Br} = 4 * 10^{-2}$ g, is: $\Delta Q_{m\,c} = 2.4$ J.

We find the value of energy transferred from the Sun to the crystals of bromine, assuming that the density of the solar energy is P = 1.4 kW / m^2. The area of the longitudinal cross section of the cylinder with a conical head is $S_{long} = d_{Br\,1} * l_t = 0.07 * 3 \approx 0.02$ cm^2. The flux density of the solar energy incident on a crystal of bromine is then equal to: $p_{Br} = 0.03$ W. It can be seen that during the flight of the crystallite $\tau_{Br} = 100$ seconds the energy required for melting the bromine crystal can be obtained from the Sun.

Besides the fact that it is necessary to melt the crystal of bromine, it has to be heated to the temperature of evaporation: $T_w = 59.2\ ^0$C and transferred to the vapor state. The energy required for the phase transition of bromine from liquid to vapor is $\Delta Q_{w\,c} = 29.5$ kJ / mole, i. e., it is about 3 times greater than the energy required for the phase transition of the solid body into the liquid state. This means that it is necessary to have the bromine crystals in the form of plates, i.e.: it is needed to develop the surface, through which the bromine crystals obtain the solar heat.

Thus, the bromine plate with a thickness of $\delta_{Br} = 7\mu$, the length of the conical part $l_{c\,p} = 0.1$ mm, and with the sizes $l_{Br} = 3.7$ x 3.7 cm, has the appropriate drag coefficient, and the plate area is 700 times greater than the area of the longitudinal cross section of the cylinder.

3.7. Slow down of bromine solid plates related with air viscosity

Since the required form of the bromine crystals is the form of thin plates with a big lateral surface, then in the subsonic velocities the bromine crystals slowing down may be very strong due to the viscosity of the air.

The equation of motion of a solid bromine plate in the air can be written as follows:

$$m_{Br}\, dV_x/dt = -F_{x2}. \tag{10}$$

the force of the resistance related with viscosity:

$$F_{x2} = \sigma S = 2\eta(dV/dz)*l_{Br}^2 = 2\eta(V/\delta)*l_{Br}^2, \tag{11}$$

where $\eta = 2 * 10^{-4}$ Poise is the viscosity of air, $l_{Br} = 3.7$ cm - side of the plate square, dV / dz - radial gradient of the longitudinal air velocity near the plate,

δ – the characteristic length of velocity changing along the radius.

Equation (10) having the viscous force of slowing down of the bromine plate in the form of (11) can be transformed into:

$$dV/V = -(2\eta/m_{Br}\ \delta)*l_{Br}^2*dt, \qquad (12)$$

which has the following solution:

$$V = V_0 exp[-(2\eta/m_{Br}\ \delta)*l_{Br}^2*t]. \qquad (13)$$

Now we can find the length being passed by the plate before its stopping in the air:

$$l_{air} = \int_0^\infty Vdt = V_0(2m_{Br}\ \delta/\eta* l_{Br}^2). \qquad (14)$$

To find the length of the above length, it is necessary to find δ which is a characteristic length of the air velocity changing near the plate.

Find it from Navies equation:

$$\partial V_x/\partial t + V_x\partial V_x/\partial x = (\eta/\rho))\partial^2 V_x/\partial z^2 + (\eta/\rho)\partial^2 V_x/\partial x^2. \qquad (15)$$

Let us see in which case the term $V_x\partial V_x / \partial x$ will be much bigger than $(\eta / \rho)\ \partial^2 V_x /\partial x^2$. Let $\partial x = l_{ch}$ –that is a characteristic length at which the velocity changes. Then $V_x\partial V_x / \partial x = V_x^2 / l_{ch}$, $(\eta / \rho)\ \partial^2 V_x / \partial x^2 = (\eta / \rho)\ V_x / l_{ch}^2$. To satisfy the condition: $V_x\partial V_x / \partial x >> (\eta / \rho)\ \partial^2 V_x/ \partial x^2$, it is necessary to fulfill $\rho V_x l_{ch} / \eta = Re >> 1$, where Re is the Reynolds number, and the ratio of Re $>>1$ in our case is certainly satisfied.

We now transform equation (15) to the following form:

$$\partial V_x/\partial t + V_x\partial V_x/\partial x = dV_x/dt = (\eta/\rho)\ d^2 V_x/dz^2 , \qquad (16)$$

where we have replaced $\partial / \partial t + V_x\partial / \partial x$ – a partial derivative with respect to time, with coordinate d / dt (the total time derivative).

Substituting the above instead of the expression for the Navies dV_x / dt, determined from the equation of motion (12),
$dV_x / dt = -(2\eta / m_{Br})*l_{Br}^2*(dV_x / dz)$, we obtain the following equation:

$$-(2\eta/m_{Br})*l_{Br}^2*(dV_x/dz)= (\eta/\rho)\, d^2V_x/dz^2, \tag{17}$$

which, after reduction by η - the viscosity of air and the above transfers, this equation changes to the following:

$$d^2V_x/dz^2 =-(2\rho/m_{Br})*l_{Br}^2*(dV_x/dz). \tag{18}$$

Replacing dV_x / dz for y we obtain:

$$d\,(\ln y) = - (2l_{Br}^2\rho/m_{Br})dz, \tag{19}$$

or

$$dV_x/dz = C_{1v}\exp(-2\rho l_{Br}^2 z/m_{Br}). \tag{20}$$

From formula (20) it is seen that the characteristic length δ, that is the change of longitudinal velocity V_x in the transverse direction z, is equal to:

$$\delta = m_{Br}/2\rho l_{Br}^2. \tag{21}$$

According to the formula (14) the slow down length of the plate in the air is equal to:

$$l_{air} = \int_0^\infty V dt = V_0(m_{Br}\delta/\eta*l_{Br}^2) = V_0\,[m_{Br}^2/\rho\eta l_{Br}^4]. \tag{22}$$

Substituting numbers into (22): $V_0= 2 * 10^4$ cm / s, $m_{Br}= 4 * 10^{-2}$ g, $\eta = 2 * 10^{-4}$ Poise, $l_{Br} = 3.7$ cm, we find the following:

$$l_{air}= 3.2*10^7 \text{ cm} = 320 \text{ km}. \tag{23}$$

It can be seen that the plate slow down caused by the viscosity of air is not important.

The estimation of the vertical velocity of the plate displacement, obtained by formula (2), has shown that this velocity is small. During the flight time of the plate over the radius of 50 kilometers, it will slightly shift in the vertical direction.

4. Operation of the equipment complex

The complex of equipment is shown in Fig. The balloon (1) raises a solid bromine gondola (2) to the required height. In the gondola there is also the appropriate cooling equipment, a centrifuge, navigation elements and power supply. The bromine plates (3) are thrown out from the gondola by means of the centrifuge over a large area. The balloon is raised to the height greater than the upper limit of the distribution of ozone in the atmosphere, shown by the curve (4). When in the atmosphere an artificial ozone hole is formed, the hard solar ultraviolet radiation reaches the Earth surface (5) and performs the required disinfection.

Conclusion

The results of this work have shown that it is possible to make the crystals reach the given point in the gaseous state by using the proper choice of the form and sizes of the bromine crystals.

The solar radiation power in the ultraviolet wavelength range can be estimated as 50 W / m^2. Thus, that the total power per area of 100 * 100 kilometers, is equal to 500 Giga Watts that is absolutely not feasible for the devices located on the surface of the Earth.

References

1. http://ru.wikipedia.org/wiki/Озоновый_слой

2. http://ru.wikipedia.org/wiki/Солнечная_радиация

3. http://elementy.ru/trefil/21185

4. I. K. Larin, Abstract dis. d. f. Th. n., Moscow, Inorganic Chemistry, 1991

5. Tables of physical quantities. Handbook ed. I. K. Kikoin,
   Moscow, Atomizdat, 1976

6. http://dic.academic.ru/dic.nsf/bse/130514/Сверхзвуковое

7. http://www.oocities.org/igor_suslov/AeroSidelnikov.pdf

8. Physical quantities. Handbook ed. I. S. Grigoriev
   and E. Z. Meilikhov, Moscow, Energoatomizdat, 1991

Two Tasks to Join Two Bodies

The article considers the transverse motion of a body having mass $m_3 = 10$ g and moving with longitudinal velocity $V_3 = 9$ km / s. It is shown that due to the correction of the motion trajectory the initial angular deviation of the body from the desired direction reduces by e times during t = 1.5 s. For the same body accelerated by means of the gas-dynamic method till the hypersonic velocity, we find conditions under which there is a solution of joining the two bodies, i.e., at a certain moment of time their coordinates coincide.

Introduction

The tasks of joining two bodies refer to the field of General Physics and can be solved by using ordinary differential equations.

I. Almost ballistic motion of bodies

Below we will consider the problem related to the field of controlled motion of bodies. We will consider the transverse motion of a body having a diameter $d_3 = 12$ mm, longitudinal velocity $V_3 = 9$ km / s and mass of $m_3 = 10$ g. We assume that angle φ between the velocity of the body and its longitudinal direction has the initial deviation $(\Delta\varphi)_{in} \approx 3 * 10^{-3}$.

The scheme of mutual arrangement of bodies in the space

Let body1be moving in a horizontal plane in vacuum with velocity V_1. In the same plane let body 2 is moving with velocity V_2 at an angle towards the direction of velocity V_1. Body 2 is moving in such away to meet body1 in point O, Fig 1.

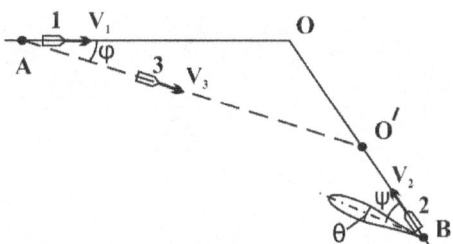

Fig.1.The scheme of the mutual arrangement of the bodies in the space

Let body (1) produce body (3) at angle φ to its velocity. The velocity of body 3 should be of such a value to meet body (2) at O' which should be located on the trajectory of body (2). This should happen before body (2) reaches point O. The velocity of body (3) is sure to consist of one projection of the velocity of body (1) on the direction of the velocity of body (3) and its own V_{31}, which is of the order $V_{31} \approx 1$ km / s. Thus, the velocity of body (3) is equal to the following:

$$V_3 = V_1 * \cos \varphi + V_{31}. \qquad (1)$$

Distance AO' should be covered by body 3 during τ_{meet} equal to the time for body 2 which passes distance BO':

$$\tau_{meet} = AO' / (V_1 * \cos \varphi + V_{31}) = BO'/V_2 \approx 3 \text{ s.} \qquad (2)$$

This condition can be satisfied (at fixed velocities V_1 and V_2 and angle φ) by selecting the proper velocity V_{31}.

We assume that body 2 is irradiating body1with the radio ray. It is clear that angle Ψ between the same velocity V_2 and the direction of body 1 will be all the same at the uniform rectilinear motion of the bodies. Meanwhile body 3 will always be on line A'B' connecting body1 and body 2. Body 3 will be irradiated with the radio ray under the same angle all the time.

The trajectory correction system

We assume that body 3 is induced to point O' on the radio ray of body 2, irradiating body 1. Fig.2. shows the structure of body 3.

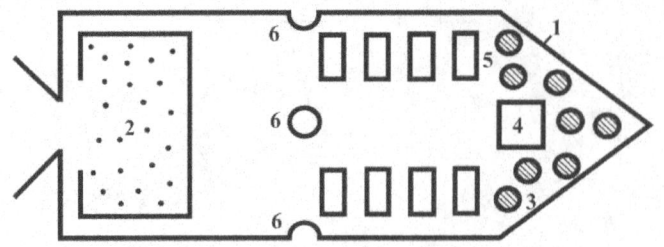

Fig. 2. Body 3 structure: (1) – housing, (2) - a jet engine, which gives longitudinal velocity to the body, (3) - multipetal receiving antenna (4) - navigation and control systems, (5) - four rows of gunpowder engines giving the transverse velocity to the body, (6) - circular hole (nozzle) through which propellant gases release.

Let the half-width of the ray irradiating the body from the distance of $l_m = 100$ km, be equal to: $r_{ray} = 10$ km. The trajectory correction system should consist of petals disposed in the head part of the body 3. Each petal is a receiver of the centimeter radiation range. After detecting the signals from each of two opposite petals, (we call them right - left and up – down), they are sent to subtraction schemes.

If body 3 moves are being placed on the axis of the radio ray, the signals are mutually compensated and there is no compensating signal in the output. Let the signal value on the axis of the radio ray be equal to: $U_{ray} = 1$ V, and at a distance of $r_{ray} = 10$ km from the radio ray, the signal drops to zero. When body 3 is not located on the radio ray axis, at the distance between the centers of the petals $\Delta r_{lobe} = 5$ mm, the difference signal will be equal to the following:

$$\Delta U_{ray} = U_{ray} * \Delta r_{lobe} / r_{ray} = 0.5 \ \mu V. \qquad (3)$$

The sensitivity of the apparatus in this case should be of the order of nano volts to confidently detect the received signals and generate a proper command for operation of a jet engine.

It is necessary to note that while body 3 is coming closer to body 2, signal $U_{ray} = 1$ V can increase by several orders of magnitude.

Gunpowder jet-engines

Let four holes with a diameter $d_{hole} = 2$ mm is located over the perimeter of body3. Under these holes let us place four guides containing the bullet sleeves filled with gunpowder. Let the diameter of the bullet sleeve be equal to the diameter of the hole: $d_{case} = d_{hole} = 2$ mm. The height of this bullet sleeve is equal to $h_{case} = 4$ mm.

Since you cannot re-energize the burning of the solid fuel jet engine, it is necessary to use another engine. This design is expected to achieve bullet sleeve by means of mechanical shifting relative to the hole. We assume that in each row there is a fixed number of 10 bullet sleeves. When the diameter of each sleeve $d_{case} = 2$ mm and the distance between the sleeves $\Delta l_{case} = 1$ mm, the total length of each of the rows will be 30 mm. It is also necessary to increase the length of body 3 by 30mm to house the used bullet sleeves since their reuse is not possible. So, taking into account the placement of the jet engine which gives the longitudinal velocity to body 3, the total length of body 3 is about 15 cm.

1. The transverse velocity value

Let us calculate the transverse velocity value of body3 engaging one bullet sleeve. We assume that the velocity of releasing the powder gases to be equal to $V_{gas} = 3$ km / s, [1].

The mass of the powder can be found from the following considerations. The volume occupied by the powder is: $V_{case} = \pi d^2_{case} * h_{case} / 4 = 10^{-2}$ cm^3. Considering the density of the powder to be equal to: $\rho_{poud} \approx 1$ g/cm^3, we find that the mass of the powder placed in the bullet sleeve, is $m_{poud} = 10$ mg.

The transverse velocity acquired by body 3, engaging one bullet sleeve, can be found from the relation:

$$V\perp = m_{poud} * V_{gas} / m_3 = 3 \text{ m/s}. \qquad (4)$$

Distance AO $' = l_{fly} = 30$ km canbe covered by body 3 during $\tau_3 = l_{fly} / V_3 = 3$ s, and it may be shifted during this time by distance $l_3 = V\perp * \tau_3 = 10$ m.

2. Horizontal flight

we consider what conditions are necessary to fulfill to preserve the horizontal level of the flight of body 3 at a distance of $l_{fly} = 10$ km. This distance is covered by body 3 during $\tau_{body} = l_{fly} / V_3 = 1$ s and under the influence of gravity, it acquires velocity $V_{grav} = g \tau_{body} = 10$ m / s, where $g = 10$ m/s^2 is the acceleration of gravity. To compensate this transverse velocity, it will be needed to use three jet engines because the operation of one jet engine gives the velocity $V\perp = m_{poud} * V_{gas} / m_3 = 3$ m / s. However, when the velocity of the body is $V_3 \approx 9$ km / s, the acceleration of the gravity will be compensated by the centrifugal force $V_3{}^2/R_E$, where $R_E = 6400$ km, the radius of the Earth.

The equation of the transverse motion

We assume the diagram of the radio ray to be equal to
$\exp [- (\Delta\Theta)^2]$, where $\Delta\Theta$ – is the deviation of the radio ray axis.
The equation of motion of body 3 in the transverse direction can be written as follows:

$$m_3 dV\perp/dt = F\perp, \qquad\qquad (5)$$

where $m_3 = 10$ g – the mass of body 3, $V\perp$ - transverse velocity, $F\perp$ - force acting in the transverse direction.

The above forces can be found from the formula $F\perp = m_{poud} * V_{gas}/\tau_0$, where $m_{poud} = 10$ mg, the mass of the gunpowder in one of the jet engines, $V_{gas} = 3$ km / s - velocity of releasing the powder gases, $\tau_0 \approx 10^{-3}$ s - response time of one jet engine [1].

We introduce the angle deviation $\Delta\varphi$, between the direction towards point O' and the longitudinal velocity of body 3. Angle $\psi = \pi / 2 - \Psi - \varphi$, is the angle between the geometrical axis of the radio ray and the direction onto body 3.

We assume that the deviation of the angle φ, is equal to $\Delta\varphi$, that is approximately equal to $\Delta\varphi = \Delta\Theta$. As we assumed above, the initial deviation of the angle should be small enough: $(\Delta\varphi)_{in} \approx 3 * 10^{-3}$. The task of the trajectory correction is to compose transverse velocity $V\perp$, at which the angle $\Delta\varphi = V\perp / V_3$ will be equal to zero. We introduce a dimensionless derivative and then the equation of the body angle can be written as follows:

$$m_b (V^2{}_3/z_0) d(\Delta\varphi)/d\tau = F\perp,$$

or

$$d(\Delta\varphi) /d\tau = m_{poud} * V_{gas} (z_0 /\tau_0)/ (m_3 V^2{}_3). \qquad\qquad (6)$$

To perform self-inducing of the body, the dimensionless force in the right side of equation (6) should depend on the magnitude of the angle deviation angle. At small angles of deviation of body 3 from the direction of point O' we can put $\Delta\varphi \approx \Delta\Theta$.

The multi petal sensor measures the derivative of function $F = \exp(-\Delta\Theta^2)$, which is equal to:

$$dF/d(\Delta\Theta) = -2(\Delta\Theta) * \exp[-(\Delta\Theta)^2]. \qquad (7)$$

Assuming the exponent to be equal to a unit, for small angles we obtain the following:

$$d(\Delta\varphi)/d\tau = -2(\Delta\varphi)* (m_{poud}* V_{gas}*(z_0/\tau_0)/ (m_3 V_3^2), \qquad (8)$$

or simply

$$d(\Delta\varphi)/d\tau = -k(\Delta\varphi), \qquad (9)$$

where: $k = 2*(m_{poud}* V_{gas}*(z_0/\tau_0)/ (m_3 V_3^2)$.

Equation (9) can be written a in the following form:

$$\Delta\varphi = (\Delta\varphi)_{in} \exp[-2*(m_{poud}* V_{gas}(z_0/\tau_0)* \tau/(m_3 V_3^2)]. \qquad (10)$$

This means that the initial deviation angle $(\Delta\varphi)$ will tend to zero, and will decrease by e times during the following period of time:

$$\tau = m_3 V_3^2/[2* m_{poud}* V_{gas}*(z_0/\tau_0)]. \qquad (11)$$

Now coming to the dimension derivative of time by replacing the variable $t = \tau * (z_0 / V_3)$, we obtain:

$$t_{corr} = m_3 V_3 \tau_0/[2* m_{poud}* V_{gas}]. \qquad (12)$$

We substitute the numbers into the formula (12) and find that reducing the initial angle of deviation from the direction of point O' will exponentially decrease with the time constant:

$$t_{corr} = (10*9*10^3 *10^{-3})/(2*10^{-2} *3*10^3) = 1.5 \text{ s}.$$

Distance AO'= l_{fly} = 30 km will be covered by body 3 during $\tau = l_{fly} / V_3 \approx 3$ s, that is by twice longer than the time constant t_{corr} = 1.5 s. It is seen that if the initial (before trajectory correction) transverse deviation Δr_{in} was of the order $\Delta r_{in} = (\Delta\varphi)_{in} * l_{fly}$ = 100 m, after correction this deviation will be by e^2 times smaller.

The obvious way to reduce the transverse deviation of body 3 - is to reduce the response time τ_0 of the jet engine.

II. Motion of a fast maneuvering body

Let body 1 move quickly maneuvering in the space. Since all the coordinates of the body maneuver independently, we can consider the motion only on one coordinate. The solution of the equation of motion with acceleration a1 of the body can be written as follows:

$$x = x_{1in} + V_1 t + a_1 t^2 /2. \qquad (13)$$

Assume the value of a1 for body 1 to be equal to: $a_1 = 3 * 10^4$ cm/s^2 or in more familiar terms, it is 30 g, where g = 10 m/s^2 is the acceleration of gravity.

We assume that body 2 accelerated by means of the gas dynamic method till hypersonic velocity [2], can change only its transverse velocity. Then the coordinates of body 2 can be written as:

$$x = x_{2\,in} + V_2 t + a_2 t^2 /2. \qquad (14)$$

Gunpowder jet engines

Let locate four holes with a diameter of d_{hole} = 10 mm over the perimeter of body 2. Under these holes let us place four guides containing the bullet sleeve filled with gunpowder. Let the diameter of the bullet sleeve be equal to the diameter of the hole $d_{case} = d_{hole}$ = 10 mm. Let the height of the bullet sleeve be equal to h_{case} = 40 mm.

Since you cannot re-energize burning of the solid fuel of the jet engine, it is necessary to use another engine. This design is expected to achieve shifting of the bullet sleeve by means of mechanical shifting relatively the hole. We assume that in each row there are 30 fixed bullet sleeves. When the diameter of each bullet sleeve d_{case} = 10 mm and the distance between the sleeves

$\Delta l_{case} = 1$ mm, the total length of each of the rows will be 310 mm. It is also necessary to increase the length of body 2 by the same length to accommodate it with the used bullet sleeves because their reuse is not possible. So, the total length of body 2 will be about 70 cm.

The magnitude of the transverse velocity

We calculate the transverse velocity of body 2 after engaging one bullet sleeve. We assume the velocity of the powder gas releasing to be equal to $V_{gas} = 3$ km / s, [1] .

The mass of the gunpowder can be found from the following considerations. The volume occupied by the gun powder is: $V_{case} = \pi d^2_{case} * h_{case} / 4 = 2.5$ cm^3. Considering the density of gun powder to be equal to $\rho_{poud} \approx 1$ g/cm^3, we find that the mass of the gun powder placed in one bullet sleeve is $m_{poud} = 2.5$ g. The total mass of the gun powder was found to be equal to: $2.5 * 4 * 30 = 300$ g. Suppose it is 50 % of the body 2 mass, i.e., its mass is $m_2 = 0.6$ kg. The transverse velocity acquired by body 2 after engaging one bullet sleeve can be found from the following relation:

$$V_\perp = m_{poud} * V_{gas} / m_2 = 12.5 \text{ m/s}.$$

Considering the response time of the gun powder engine to be equal to $\tau_0 = 1$ms [1], we find that the acceleration of body2 is equal to: $a_{22} = V_2/\tau_0 = 1.25 * 10^6$ cm/s^2. Note that this impulse acceleration by 40 times exceeds the acceleration of body 1 equal to: $a_1 = 3 * 10^4$ cm/s^2.

However, not the pulsed but average acceleration is important which is received by the body for a certain period of time. Assuming that the repetition rate of the engine is $F = 100$ Hz, we find that the time interval between the operation of the engine is 10 ms, i.e., the average acceleration in this case is: $a_2 = 1.25 * 10^5$ cm / s^2, and it is 4 times higher than the acceleration of body1 equal to: $a_1 = 3 * 10^4$ cm/s^2.

We require that the coordinates of body 1 and body 2 coincide in a certain moment of time. Then we obtain a quadratic equation in dependence on time:

$$x_{1in} + V_1 t + a_1 t^2 /2 = x_{2in} + V_2 t + a_2 t^2/2, \qquad (15)$$

$$t^2 + 2(V_2 - V_1) t/(a_2 - a_1) + 2(x_{2in} - x_{1in})/ (a_2 - a_1) = 0. \qquad (16)$$

For the equation to have real roots, it is necessary to satisfy the following condition:

$$(V_2 - V_1)^2 > 2*(x_{2in} - x_{1in})*(a_2 - a_1). \tag{17}$$

We introduce the notation: $\Delta x_{in} = (x_{2in} - x_{1in})$ —which is the difference between the initial values of the coordinates of body 1 and body 2 and $\Delta t_{reac} = V / a$ - response time of body 2 due to the change of the coordinate of body 1. Equation (17) can be written as follows:

$$a_2 > a_1 + [2\Delta x_{in}*(a_2 - a_1)/ \Delta t^2_{reac}]^{1/2}. \tag{18}$$

In our case, for $a_2 \gg a_1$ the above expression (18) can be rewritten as:

$$a_2 > 2 \Delta x_{in}/ \Delta t^2_{reac}. \tag{19}$$

Actually, expression (19) has been reduced to a trivial expression:

$$\Delta x_{in} < a_2*\Delta t^2_{reac}/2, \tag{20}$$

- to the restriction on the initial difference of the coordinates. For the considered case the response time of all the engines in the same row is equal to Δt_{reac} power: $\Delta t_{reac} = 0.3$ s and the initial difference found from the relation (20) must be less than $\Delta x_{in} < 60$ m.

Conclusion

Taking into account all the coordinates considered above, we have to conclude that to make two bodies moving in the space coincide, it is necessary to have all the three coordinates coincide.

Literature

1. http://ru.wikipedia.org/wiki/Ракетный_двигатель

2. S. N. Dolya, RF Application for the Grant of a Patent for Invention № 2014106015, http://arxiv.org/ftp/arxiv/papers/1403/1403.4541.pdf

Concentrator of elastic waves

This article is dedicated to an opportunity of concentrating elastic waves in the iron and water cones on the square of the cone vertex of the order of 1 cm². The square of the base of the cone is equal to1 m², its height - 1 m. The calculations assume that the cone hexogen network lying in the cone basis explodes during the time of 1 μs and causes an explosive wave converging to the vertex of the cone. It is shown that this explosive wave can accelerate the body having a mass of 3 g to speed V = 5 km / s.

1. Physical motivation of the task

Suppose there is an iron cone, which is a part of a ball with radius $r_0 = 1$ m and a basic square of $S_{cone} = 1$ m². On a spherical base of the cone let us place a grid with a cell size h = 3 cm, in whose nodes there are hexogen balls with a diameter of $d_{hex} = 1$ cm. The volume of the ball $V_{hex} = \pi d^3_{hex} / 6 \approx d^3_{hex} / 2$ or 0.5 cm³. At a density of RDX [1] $\rho_{hex} \approx 1.8$ g / cm³, the mass of each ball is about 1g. Obviously, that the total number of balls placed on the base of the cone is 10^3, and their total mass $m_{hex} = 1$ kg.

At the simultaneous undermining of all the balls, the released energy will be of the order of $\Delta Q_{hex} \approx 6$ MJ. The velocity of the detonation wave in RDX is $V_{det. wave} = 8.36$ km / s, so that we can assume that the time of the energy release will be $\Delta \tau_{hex} = 1$ μs. The power of the energy release will be equal to the following:

$$W_{hex} = \tfrac{1}{2}\Delta Q_{hex}/\Delta \tau_{hex} = 3*10^{12} \text{ W,} \qquad (1)$$

where we have taken into account that only half of the energy released in the explosion will get inside the cone.

The density of energy release power that is the intensity of the wave (pulse) near the base of the cone is equal to:

$$I_1 = W_{hex}/ S_{cone} = 3*10^{12} \text{ W/m}^2. \qquad (2)$$

In the region with an area of $S_{vert} = 1$ cm² at the cone vertex there will be concentration of the explosion energy, the intensity of the pulse there will be bigger if to take the square of the base and the square of the vertex:

$$I_2 = I_1 S_{cone}/ S_{vert} = 3*10^{16} \text{ W/m}^2. \qquad (3)$$

95

2. The amplitude of oscillations

The pressure in the sonic wave is related with the sound intensity by the following ratio:

$$P_{sound\ Fe} = (I_2 * \rho_{Fe} * V_{sound\ Fe})^{1/2}, \qquad (4)$$

where I_2 – the sound intensity near the vertex, $\rho_{Fe} = 9 * 10^3$ kg / m^3 – the density of iron, $V_{sound\ Fe} \approx 6$ km / s – the velocity of longitudinal elastic waves in iron [2], p. 86.

Substituting numbers into the formula (4), we find that the pulse pressure near the vertex of the cone is equal to: $P_{Fe} = 1.3 * 10^{12}$ Pa.

Now we find the rate of displacement of atoms V_{disp} in this pulse from the following relation:

$$P_{Fe} = \rho_{Fe} * V^2_{disp}/2, \qquad (5)$$

where

$$V_{disp} = (2 * P_{Fe}/ \rho_{Fe})^{1/2} = 1.7 * 10^4 \text{ m/s.} \qquad (6)$$

We have obtained the velocity of the atom displacements to be equal to $V_{disp} = 17$ km / s, which is by twice larger than the first space velocity $V_{1\ space} = 8$ km /s.

This shows that if to place a small body in the vertex of such a cone, a significant portion of the explosion energy will be transferred to this body which can be launched into space.

It can be seen that the iron cone will be damaged with each explosion and its reuse is hardly possible.

3. Water cone

Let us consider an opportunity of concentrating the energy of elastic oscillations by using the water cone having the same sizes as the iron one. Water may be placed into the Mylar or rubber shell. The volume of the cone is

$V_{cone} = (1/3) r^3_0 = 0.3$ m^3. When the density of water $\rho_{water} = 10^3$ kg / m^3, the mass of water in this cone, will be approximately equal to $m_{water} = 300$ kg. The cone should be positioned inside a hard steel trunk to prevent the damage of the water cone in the transverse direction.

We find the pressure in the sonic wave near the cone vertex:

$$P_{sound\ water} = (I_2 * \rho_{water} * V_{sound\ water})^{1/2}. \tag{7}$$

Substituting numbers into the formula (7), where $\rho_{water} = 10^3$ kg / m^3, $V_{sound\ water} = 1.5$ km / s, we find that $P_{sound\ water} = 2 * 10^{11}$ Pa.

The rate of displacement of water molecules in the pulse is equal to:

$$V_{disp} = (2 * P_{water} / \rho_{water})^{1/2} = 1.4 * 10^4 \text{ m/s}, \tag{9}$$

it also turned out to be greater than the first space velocity.

4. Non-linear theory

It is clear that at such high pressures of $P_{sound\ water} = 2 * 10^{11}$ Pa, the density of water will change, and together with it the other parameters will also change. The system of equations describing the state of condensed water can be represented as follows [3]:

$$\rho / \rho_0 = V_s / (V_s - V_d); \tag{10}$$

$$P_s + \rho(V_s - V_d)^2 = \rho V_s^2; \tag{11}$$

$$P_s = A[(\rho / \rho_0)^n - 1]; \tag{12}$$

where ρ_0 and ρ are the density of water in front and behind the explosive wave front, V_s – the velocity of propagation of the explosive wave front, V_d - the mass velocity behind the explosive wave front, $A = 3 * 10^8$ Pa, n = 7- 8. Equation (10) expresses the law of conservation of mass, equation (11) - the law of conservation of momentum, and the equation (12) is the equation of state of water in the compressed form.

Let $\rho / \rho_0 = 3$, $V_s = 15$ km / s. Then the pressure in the sound wave will be more than the linear pressure (equation (7)) by $(30)^{1/2}$ times, i.e. it is equal

97

to $P_s \approx 10^{12}$ Pa. From (12) we find that P_s for $\rho / \rho_0 = 3$ and $n = 8$ is equal to $P_s = 2 * 10^{12}$, which can be considered to be a good agreement between the above.

From equation (10) we find that in this case that the mass velocity behind the explosive wave front will be equal to $V_d = 10$ km / s.

We substitute these values in the formula (11). We find:

$$10^{12} + 7.5*10^{10} = 6.75*10^{11}.$$

That may be also considered a rather good agreement.

5. Energy transfer to the body

Suppose that in the vicinity of the cone vertex there is is a physical body with a cross-section of 1 cm^2 and a mass of $m_b = 3$g. From the law of conservation of momentum it follows that in this case the velocity of $V_b = 5$ km / s will be transferred to the body.

The kinetic energy of the body in this case will be equal to:

$$m_b V_b^2/2 = 37.5 \text{ kJ}, \tag{13}$$

and the coefficient of energy transfer from the explosion to the physical body will be equal to 37 kJ / 6 MJ $\approx 0.6\%$.

Let us consider the maximum lifting height and range of the flight of the body, released at the angle of 45^0 to the horizon, taking into account the air resistance. The velocity of the body due to this air resistance decreases with time of flight according to the law [4]:

$$V(t) = V_0/(1+\rho_0*\exp(-z/H_0)*C_x*S_{tr}* V_0 t/2m), \tag{14}$$

where V_0 - initial velocity, $\rho_0 = 1.3 * 10^{-3}$ g / cm^3 - the density of the air near the Earth surface, $H_0 = 7$ km – a barometric coefficient, m - mass of the body, C_x - drag coefficient of the body, S_{tr} - cross-section of the body, z - lifting height of the body.

Without the air resistance the formulae of the flight range and maximum lifting height are as follows:

$$S_{max} = 2V^2_0 * \sin\Theta * \cos\Theta / g, \qquad\qquad (15)$$

$$H_{max} = V^2_0 * \sin\Theta * \cos\Theta / 2g, \qquad\qquad (16)$$

where $g = 10^{-2}$ km / s^2. Then for $V_0 = 5$ km / s and the angle of inclination to the horizontal velocity of $\Theta = 45^0$, we would have obtained the maximum flight range $S_{max} = 2500$ km and maximum lifting height $H_{max} = 625$ km.

However, due to the air resistance, the velocity V (t) decreases according to the law given by the formula (14). The density of the atmosphere decreases exponentially according to the barometric formula. At the height of H = 7 km it is already less by e times than that at the Earth surface, where e = 2.72 – the base of natural logarithms.

If you use a bullet, whose diameter is much less than the diameter of the trunk, then for this bullet it is possible to have a small drag coefficient C_x.

The reason why we can make a small drag coefficient lies in its dependence on the velocity for a sharp cone. At hypersonic velocities the drug coefficient does not depend on the velocity of the body and becomes equal to a constant value:

$$C_x = \Theta^2_{vert}, \qquad\qquad (17)$$

where Θ_{vert} - the angle at the vertex of the cone. Fo the body of an elongated form, which has the ratio of the body length l_{body} to its diameter d_{body} much greater than the unity $l_{body} / d_{body} >> 1$, the angle at the vertex of the vertex can be made sufficiently small, for example, $\Theta_{vert} = d_{body} / l_{body} = 0.1$. Then the drag coefficient $C_x = 10^{-2}$, and it can be expected that the decrease of the body velocity while its crossing the atmosphere will be small. After crossing the atmosphere by the body you can use formulae (15) and (16) by substituting the appropriate velocity value into the formulae, which is obtained after the body crossing of the atmosphere.

Here it is necessary to take into account that the body rises into the atmosphere not vertically, but at the angle of 45^0. This can be taken into account by dividing the second term in the denominator by $\sin \Theta$, i.e., multiplying it in this case by 1.41.

For the body with a diameter $d_{body} = 3$ mm, the vertex angle $\Theta_{vert} = 0.1$, the mass $m_{body} = 3$ g, after three seconds of the flight we obtain: $V = 3.4$ km / s. The average velocity of the body at this distance will be equal to $\bar{V} = 4.2$ km / s, the lifting height during this lifting time will be 9 km. We assume that further the body will fly without air resistance. Then, according to (15), (16) the body with velocity $V = 3.4$ km / s will fly $S_{max1} = 1100$ km and rise to the height of $H_{max1} = 290$ km.

Conclusion

It is clear that though this concentrator of elastic waves allows it to transfer the hypersonic velocity to a small body, but the efficiency of energy transfer is small.

References

1. https://ru.wikipedia.org/wiki/Гексоген

2. Tables of physical quantities. Handbook ed. I. K. Kikoin, Moscow, Atomizdat, 1976

3. A. G. Russian, V. I. Oreshkin, A. Yu. Lyubetskii et al. Study electrical explosion of conductors in the high pressure of the converging shock wave, Zh. Tech. Phys, t.77, issue 5, p. 35, 2007, http://journals.ioffe.ru/jtf/2007/05/p35-40.pdf

4. http://arxiv.org/ftp/arxiv/papers/1403/1403.4541.pdf